A Whistle-Stop Tour of Statistics

D0165709

A Whistle-Stop Tour of Statistics

Brian S. Everitt

CRC Press
Taylor & Francis Group
Boca Raton   London   New York

CRC Press is an imprint of the
Taylor & Francis Group an **informa** business

A CHAPMAN & HALL BOOK

CRC Press
Taylor & Francis Group
6000 Broken Sound Parkway NW, Suite 300
Boca Raton, FL 33487-2742

© 2012 by Taylor & Francis Group, LLC
CRC Press is an imprint of Taylor & Francis Group, an Informa business

Printed in the United States of America on acid-free paper
Version Date: 20111025

International Standard Book Number: 978-1-4398-7748-7 (Paperback)

Visit the Taylor & Francis Web site at
http://www.taylorandfrancis.com

and the CRC Press Web site at
http://www.crcpress.com

To all my long-suffering doubles partners at the Edward Alleyn Tennis Club

Contents

Preface

According to my *Penguin English Dictionary*, whistle-stop, used before a noun, means 'consisting of brief stops in several places' and this whistle-stop tour of statistics does just that, with the ten 'stops' being ten major areas of statistics. In *A Whistle-Stop Tour of Statistics* quintessential accounts of the topics which are the subject of each part are given with the summaries at the end of each part collecting together the most important formulae, etc. The book is intended as a quick source of reference and as an *aide-memoir* for students taking A-level, undergraduate or postgraduate statistics courses. The numerous examples included in the 'tour' may also be helpful to instructors on such courses by providing their students with small data sets with which to work. The book was partially suggested by the two excellent 'Companions' by A.C. Fischer-Cripps:

Fischer-Cripps, A.C. (2003) *The Physics Companion*, Taylor & Francis, New York.
Fischer-Cripps, A.C. (2005) *The Mathematics Companion*, Taylor & Francis, New York.

Some Basics and Describing Data

1

1.1 POPULATION, SAMPLES AND VARIABLES

Population: The set of *all* 'individuals' (often humans, but not always so) that are of interest in a study; for example, all men born in Essex in 1944, all women between the ages of 20 years and 60 years living in San Francisco, all companies in Paris with fewer than 20 employees.

Sample: A subset of the 'individuals' in a population; for example, 1000 men born in Essex in 1944, 100 women between the ages of 20 and 60 living in San Francisco, 10 companies in Paris with fewer than 20 employees. Samples may be taken in various ways but in most cases we assume *random sampling,* meaning that each individual in the population has the same probability of being sampled.

Data: Information collected on members of the sample by measuring, counting or observing the value (or values) of some *variable* (variables) on each sample member where a variable is any quantity that may vary from individual to individual; for example, weight of each of the 1000 Essex men, number of sexual partners for each of the 100 San Francisco women, annual turnover for each of the 10 Paris companies.

Nearly all statistical analysis is based on the principle that one collects data on the members in the sample and uses this information to make some *inferences* about the population of interest. Samples are needed because it is rarely possible to study the whole population. The relation between sample and population is subject to uncertainty and we use probability concepts to quantify this uncertainty (see Chapters 2, 3 and 4).

1.2 TYPES OF VARIABLES

Four types of variable may be distinguished:

Nominal (categorical) variables: Variables that allow classification with respect to some property; examples are marital status, sex and blood group. The categories of a nominal scale variable have no logical order; numbers may be assigned to categories but merely as convenient labels for these categories.

Ordinal variables: Variables that have one additional property over a nominal scale variable, namely, a logical ordering of categories; now the numbers assigned to categories indicate something about the amount of a characteristic possessed but not about the differences in the amount. Examples of such variables are ratings of anxiety and depression and assessments of IQ.

Interval variables: Variables possessing a further property over ordinal scales, and that is that equal differences on any part of the scale reflect equal differences in the characteristic being measured. The zero point for such scales does not represent the absence of the characteristic the scale is used to measure. Examples of interval scale variables are temperatures measured on the Celsius (C) or Fahrenheit (F) scales.

Ratio variables: Variables that differ from interval scale variables in having a true zero point that represents the absence of the characteristic measured by the scale. Examples are temperature on the Kelvin (K) scale and weight.

The different types of measurement scales *may* require the application of different types of statistical methods for valid conclusions to be made.

1.3 TABULATING AND GRAPHING DATA: FREQUENCY DISTRIBUTIONS, HISTOGRAMS AND DOTPLOTS

The construction of informative *tables* is often one of the first steps in trying to understand a data set.

EXAMPLE 1.1

Eye colour collected on a sample of 22,361 children in Aberdeen, Scotland. Eye colour is a categorical variable. How can we usefully tabulate the data?

SOLUTION

Simply count the number of children in each category of the eye colour variable:

	EYE COLOUR			
	BLUE	LIGHT	MEDIUM	DARK
Count of children	2978	6697	7511	5175
Percentages	13.3	29.9	33.6	23.1

NB: If only percentages are given, the size of the sample on which they are based must also be quoted.

EXAMPLE 1.2

The starting positions of a sample of 144 winners of eight-horse horse races on circular tracks were recorded where starting position is numbered 1 to 8 with position 1 being closest to the rail on the inside of the track. So the original data consisted of a series of 144 numbers from 1 to 8, i.e., 1,1,3,2,4,......4,3,8. What type of variable is starting position and how can the data be tabulated to make them more transparent?

SOLUTION

If the starting positions are equally spaced out from the fence, then starting position is an interval variable but one that is discrete rather than continuous. If the starting positions are not equally spaced out from the fence then the variable is ordinal only. To tabulate the data construct a table giving the number of winners in each starting position.

	STARTING POSITION							
	1	2	3	4	5	6	7	8
Number of wins	29	19	18	25	17	10	15	11

EXAMPLE 1.3

The heights in millimetres of a sample of 169 men have been collected. What is a useful way of tabulating them?

SOLUTION

Here we can count the number (frequency) of men falling into each of a number of intervals for height to give a *frequency distribution table*.

CLASS INTERVAL	FREQUENCY
1550–1599	5
1600–1649	12
1650–1699	36
1700–1749	55
1750–1799	35
1800–1849	16
1850–1899	9
1900–1949	1

Tables can often be usefully represented by various graphics:

Bar chart: A graphical representation of data classified into a number of categories. Equal-width rectangular bars are constructed over each category with height equal to the observed frequency of the category. See Figure 1.1 for an example. Bar charts are often used but it is doubtful that they provide any advantage over the corresponding table of frequencies.

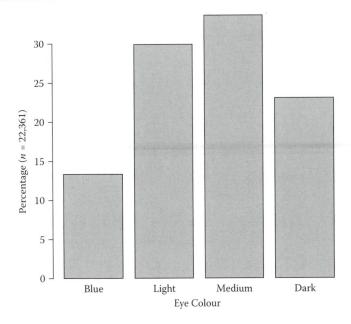

FIGURE 1.1 Bar chart of eye colour counts.

Pie chart: A widely used graphical technique for representing relative frequencies associated with the observed values of a categorical variable. The chart consists of a circle divided into sectors whose sizes are proportional to the quantities (often percentages) that they represent. An example is given in Figure 1.2. Such graphics are popular in the media but have little advantage over the tabulated data, particularly when the number of categories is small.

Dotplot: A graphic with horizontal line for each category of a categorical variable and a dot on each giving either the category frequency or the value of some other numerical quantity associated with the category. Examples are shown in Figures 1.3 and 1.4. Dotplots are generally more effective displays than both bar charts and pie charts.

Histogram: A graphical representation of a frequency distribution table in which class frequencies are represented by the areas of rectangles centred on the class interval. If the latter are all of equal length then the heights of the rectangles are proportional to the observed class frequencies. An example is shown in Figure 1.5.

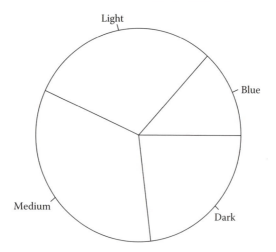

FIGURE 1.2   Pie chart of eye colour frequencies.

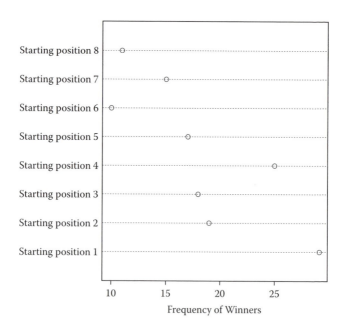

FIGURE 1.3   Dotplot of horse race winner frequency by starting position.

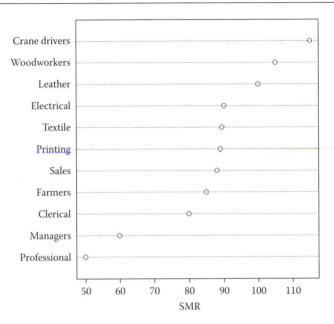

FIGURE 1.4 Dotplot of standardized mortality rates (SMR) for lung cancer in several occupational groups.

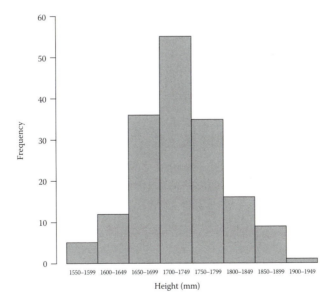

FIGURE 1.5 Histogram of heights of 169 men.

1.4 SUMMARIZING DATA: MEAN, VARIANCE AND RANGE

More concise summaries of data than tables and graphs are obtained by calculating a small number of numerical characteristics of the data; such numerical characteristics are termed *statistics* of the data or sample. Emphasis here will be on statistics for interval or ratio scale variables.

1.4.1 Measures of Central Tendency

Mean: Given a sample of n values of a variable, $x_1, x_2, x_3, \ldots, x_n$, the mean value of the variable $\bar{x}$ is

$$\bar{x} = \frac{\sum_{i=1}^{n} x_i}{n}$$

If the variable values are summarized in a frequency distribution table with k classes with midpoints $x_1, x_2, \ldots, x_k$ and frequencies $f_1, f_2, \ldots, f_k$ then the mean is calculated as

$$\bar{x} = \frac{\sum_{i=1}^{k} f_i x_i}{\sum_{i=1}^{k} f_i}$$

Median: For a sample of variable values that are *skewed* (lack of symmetry in the frequency distribution of the data) the mean may not be a useful measure of central tendency; in such cases the *median* may be more useful. The median is simply the value in the ranked variable values that divides the sample into two parts each with 50% of the variable values. When there is an odd number of observations the median is the middle value; when the number of observations is even the median is calculated as the average of the two central values. For symmetric data the sample mean and median will be approximately equal.

Mode: The most frequently occurring value in a sample.

EXAMPLE 1.4

What are the mean and the median for the following two data sets?

Set1: 3.1,5.2,2.1,6.3,7.8,4.1,4.0,8.8,0.6
Set2: 1.1,3.2,4.2,0.7,5.0,5.5,4.6,9.0,15.0,18.0

SOLUTION

Set1: Mean = 4.67, median = 4.10
Set2: Mean = 6.63, median = 4.80

1.4.2 Measures of Variability

Variance: Given a sample of n values of a variable, $x_1, x_2, ..., x_n$, the variance s^2 is given by

$$s^2 = \frac{1}{n} \sum_{i=1}^{n} (x_i - \bar{x})^2$$

(NB: The variance, s^2, is often defined with a denominator of $n-1$ rather than n for reasons that will be made clear in Chapter 3.)

Standard deviation: The square root of the variance, gives a measure of variability in the same units as the original variable values.

Range: The difference between the largest and variable values in the sample. Quick to calculate but the range is generally not a very useful measure of variability.

Inter-quartile range: The difference between the third and first *quartiles* of a sample of variable values where quartiles are the values that divide the sample into four sub-sets each containing an equal number of observations. More useful than the range because it is less affected by extreme observations very different from the bulk of the values in the sample; observations often referred to as *outliers*.

EXAMPLE 1.5

What are the variance, standard deviation, range and inter-quartile range for the following sample of suicide rates per million for 11 cities in the United States?

CITY	SUICIDE RATE
New York	72
Los Angeles	224
Chicago	82
Philadelphia	92
Detroit	104
Boston	71
San Francisco	235
Washington	81
Pittsburgh	86
St. Louis	102
Cleveland	104

SOLUTION

Variance = 3407.9, standard deviation = 58.4, range = 164, 1st quartile = 81.5, 3rd quartile = 104, inter-quartile range = 22.5.

The extreme suicide rates for San Francisco and Los Angeles (what's so bad about this cities?) tend to distort all the measures of variability except the inter-quartile range.

The quartiles of a data set divide the data into four parts with equal numbers of observations. The quartiles are the 25th, 50th and 75th *percentiles* of the data where the *P*th percentile of a sample of n observations is that value of the variable with rank $(P/100)(1+n)$; if this is not an integer, it is rounded to the nearest half rank.

1.5 COMPARING DATA FROM DIFFERENT GROUPS USING SUMMARY STATISTICS AND BOXPLOTS

Many studies involve the comparison of a sample of values on some variable of interest in different groups, for example, in men and women, in different countries, in different medical diagnostic groups. For example, Table 1.1 (known as a *contingency table*) allows a comparison of the smoking habits of people in different occupations.

TABLE 1.1   Cross-classification of smoking habits and occupation

| SMOKING CHARACTERISTIC | OCCUPATION (PERCENTAGES SHOWN) | | | TOTAL |
	BLUE COLLAR	PROFESSIONAL	OTHER	TOTAL SAMPLE SIZE
DAILY AVERAGE				
20+	44.1	6.2	49.7	8951
10–19	42.7	5.8	51.5	2589
1–9	38.6	7.6	53.8	1572
Former smoker	30.9	11.0	58.1	8509
Never smoked	30.5	11.8	58.7	9694

The table demonstrates that there is a far smaller percentage of professional workers who are heavy smokers than for other types of worker.

Comparisons of grouped data can also be in terms of summary statistics and graphics. For example, Table 1.2 gives the weights (in grammes) of 10 items as measured on two kitchen scales and Table 1.3 gives the values of various summary statistics for the two weighing machines

Boxplot (box-and-whisker plot): A useful graphic for comparing data sets which displays the distribution of the bulk of the observations and any extreme values. The graphic is based on the *five number summary of the data,* i.e., the minimum value, lower quartile, median, upper quartile and maximum value. For the weights measured by two scales the boxplots are shown in Figure 1.6. The plot clearly shows the similarity of most characteristics of the two samples of weights.

TABLE 1.2   Weights in grammes of 10 items as measured by two kitchen scales

ITEM	SCALE A	SCALE B
1	300	320
2	190	190
3	80	90
4	20	50
5	200	220
6	550	550
7	400	410
8	610	600
9	740	760
10	1040	1080

TABLE 1.3 Summary statistics for the weights provided by two kitchen scales

	MEAN	STANDARD DEVIATION	RANGE	INTER-QUARTILE RANGE
Scale A	413	321.6	1020	402.5
Scale B	427	324.7	1030	390.0

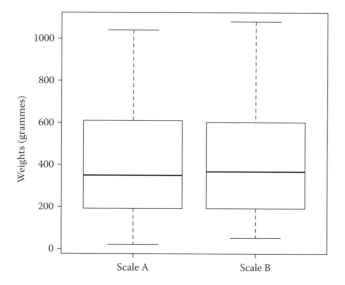

FIGURE 1.6 Boxplots for weights as measured by two kitchen scales.

1.6 RELATIONSHIP BETWEEN TWO VARIABLES, SCATTERPLOTS AND CORRELATION COEFFICIENTS

Univariate data: Data containing the values of a single variable for a sample of individuals.

Bivariate data: Data with the values of two variables for each member of the sample. Assessing the *relationship* between the pair of variables becomes important.

Scatter diagram (scatterplot): Simply a graph in which the values of one variable are plotted against those of the other; a very useful graphic for studying the relationship between two variables.

EXAMPLE 1.6

Construct a scatterplot of suicide rates for various cities in the United States (see Example 1.5) and the percentage unemployed for the same cities given below.

CITY (%UNEMPLOYED)										
NY	LA	CHI.	PHIL.	DET.	BOS.	SF	WASH.	PITTS.	ST.L.	CLE.
3.0	4.7	3.0	3.2	3.8	2.5	4.8	2.7	4.4	3.1	3.5

SOLUTION

A scatterplot of the two variables is shown in Figure 1.7. The plot shows that there is some relationship between %unemployed and suicide rates with higher values of the former being loosely associated with higher values of the latter; the suicide rates for SF and LA are very different from the rates for the other cities, and their unemployment percentages are a little higher than the percentages of the other cities.

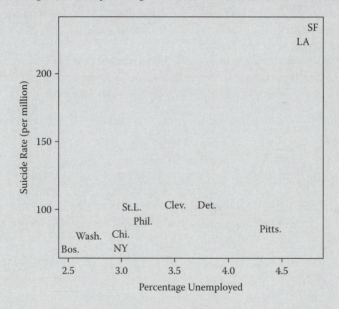

FIGURE 1.7 Scatterplot of suicide rate vs. %unemployed for 11 cities in the United States.

Sample correlation: The sample correlation is a measure of how closely the points in a scatter plot of the sample values of two variables lie to a straight line. The relationship between two variables can often be usefully quantified by calculating a measure of their correlation, i.e., a measure of the strength *and* the direction of the *linear relationship* between two random variables.

Pearson's correlation coefficient (product moment correlation coefficient): The most commonly used correlation coefficient; usually denoted by r and for n pairs of variable values $(x_1, y_1), (x_2, y_2), \ldots, (x_n, y_n)$ given by

$$r = \frac{\sum_{i=1}^{n}(x_i - \bar{x})(y_i - \bar{y})}{\sqrt{\sum_{i=1}^{n}(x_i - \bar{x})^2 \sum_{i=1}^{n}(y_i - \bar{y})^2}}$$

The coefficient r can take values between -1 and 1 with the numerical value indicating the strength of the linear relationship and the sign indicating the direction of the relationship. The plots in Figure 1.8 show scatter plots of several data sets and the associated values of r. The plots in Figure 1.8 (g) and (h) demonstrate that r is only useful for assessing the strength of the *linear* relationship between two variables.

Spearman's rho: A correlation coefficient that uses only the *rankings* of the observations. If the ranked values of the two variables for a sample of individuals are a_i and b_i with $d_i = a_i - b_i$, then Spearman's rho is given by

$$\text{rho} = 1 - \frac{6\sum_{i=1}^{n} d_i^2}{n^3 - n}$$

For the suicide and %unemployed variables in 11 cities in the United States given earlier, Pearson's correlation coefficient is 0.81 and Spearman's rho is 0.88.

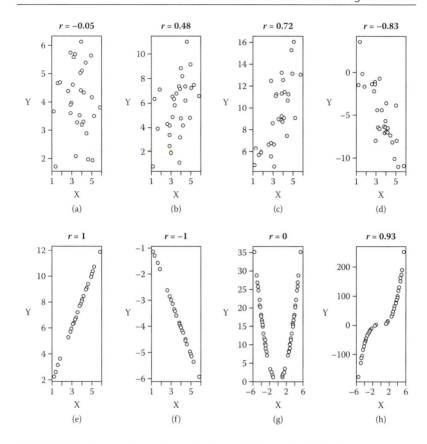

FIGURE 1.8   Scatterplots and correlation coefficients for eight data sets.

1.7 TYPES OF STUDIES

The four main types of study used in scientific research are surveys, experiments, observational studies and quasi-experiments.

- *Surveys*: Asking questions has been found to be a remarkably efficient way to obtain information from and about people whether via written questionnaires, in-person interviews, telephone conversations or the Internet. But non-response can cause problems and the questions asked need to elicit accurate responses. One of the most famous surveys of the 20th century was that conducted by Alfred Charles Kinsey into human sexual behaviour.

• *Experiments:* The essential feature of an experiment is the large degree of control in the hands of the experimenter which can, in general, allow unambiguous conclusions to be made about some treatment or intervention of interest. The *clinical trial* is the most common form of experiment with groups of participants formed by *randomization.*

• *Observational:* Studies in which the researcher does not have the same amount of control as in an experiment; in particular, people cannot be assigned at random to receive the procedures or treatments whose effect it is desired to discover, instead members of naturally occurring groups are studied. An example for which an observational study rather than an experiment is necessary is in an investigation of the relationship between smoking and blood pressure; people cannot be randomly assigned to be 'smokers' or 'non-smokers' but the blood pressure of people who smoke and those who do not can be recorded and compared. Observational studies are less powerful than experiments because there is more ambiguity in the interpretation of results. There are various types of observational studies, for example, *case-control studies*, *cross-sectional studies* and *cohort (prospective) studies*; the essential features of each type are illustrated in Figure 1.9

• *Quasi-experiments:* Such studies resemble experiments but are weak on some of the characteristics; in particular, the ability to manipulate the groups to be compared is not under the investigator's control. But interventions can be applied to naturally occurring groups.

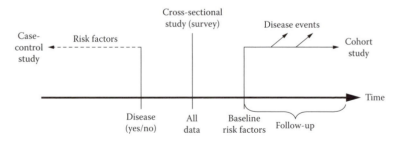

FIGURE 1.9 Schematic comparison of the three major types of observational studies.

An example is provided by a study investigating the effectiveness of different methods of teaching mathematics, in which the different methods are given to the members of different classes in the school.

1.8 SUMMARY

Population: The complete set of individuals of interest.

Sample: A sub-set of the individuals that make up a population.

Variable types: Nominal/categorical, ordinal, interval, ratio.

Bar chart: A graphic for displaying the frequencies of the categories of a categorical variable.

Pie chart: A graphic for displaying the relative frequencies associated with the observed values of a categorical variable.

Dotplot: A graphic for displaying quantitative data which are labelled in some way.

Histogram: A graphical representation of a frequency distribution table in which class frequencies are represented by the areas of rectangles centred on the class interval.

$$\text{Mean: } \bar{x} = \frac{\sum_{i=1}^{n} x_i}{n}$$

$$\text{Variance: } s^2 = \frac{1}{n} \sum_{i=1}^{n} (x_i - \bar{x})^2$$

Range: $x_{max} - x_{min}$.

Inter-quartile range: Difference between the first and third quartiles of a sample.

Percentile: The Pth percentile of a sample of n observations is that value of the variable with rank $(P/100)(1+n)$; if this is not an integer, it is rounded to the nearest half rank.

Boxplot: A useful graphic for comparing the characteristics of samples of variable values from different populations.

Scatterplot: A graphic displaying how two variables are related.

Pearson's correlation coefficient: $r = \dfrac{\displaystyle\sum_{i=1}^{n}(x_i - \bar{x})(y_i - \bar{y})}{\sqrt{\displaystyle\sum_{i=1}^{n}(x_i - \bar{x})^2 \sum_{i=1}^{n}(y_i - \bar{y})^2}}$

Spearman's rho: Pearson's correlation coefficient based on the ranks of the two variables.

Types of study: Experiments, e.g., clinical trials; observational, e.g., case-control studies, quasi-experimental.

SUGGESTED READING

Altman, DG (1991) *Practical Statistics for Medical Research*, Chapman & Hall, London.

Daly, F, Hand, DJ, Jones, MC, Lunn, AD and McConway, KJ (1995) *Statistics*, Addison Wesley, Wokingham.

Everitt, BS and Palmer, CR (2010) *Encyclopaedic Companion to Medical Statistics*, Wiley, Chichester.

Freedman, D, Pisari, R and Purves, R (2007), *Statistics*, 4th Edition, W.H. Norton and Co., New York.

Probability

2

2.1 PROBABILITY

Events: The results of experiments or observations. *Compound* and *simple events* need to be distinguished. For example, consider the age of a person, x. Each particular value of x represents a simple event, whereas the statement that a person is in her fifties describes the compound event that x lies between 50 and 60. Every compound event can be decomposed into simple events; a compound event is an aggregate of certain simple events.

Probability: A measure of that event's uncertainty. By convention probabilities are measured on a scale from zero to one, with zero corresponding to an event that is impossible and one to an event that is certain.

Classical definition of probability: If there are n equally likely possibilities of which s result in the event of interest (often referred to as a *success*) then the probability of the event, p, is given by

$$p = s/n$$

with the probability that the event does not happen (often denoted a *failure*) being $1 - p$.

EXAMPLE 2.1

What is the probability of a year, which is not a leap year, having 53 Sundays?

SOLUTION

A non-leap year of 365 days consists of 52 complete weeks and 1 day over. This odd day may be any one of the 7 days of the week and there is nothing to make one more likely than another. Only one will lead to the result that the year will have 53 Sundays; consequently, the probability that the year has 53 Sundays is simply 1/7.

Alternative definition of probability: The limit of the relative frequency of a success in a large number of trials, n, as n tends to infinity; explicitly,

$$p = \lim_{n \to \infty} \frac{s}{n}$$

where s is the number of successes in the n trials.

EXAMPLE 2.2

Suppose 1,000,000 children are born in a year, with 501,000 of them being boys. The relative frequency of boys is $501,000/1,000,000 = 0.501$. Here the denominator is large and we can justifiably claim that the probability of a male child is 0.501.

Now suppose that in a small village somewhere, 10 children are born in a year of which 7 are boys. The relative frequency of male babies is $7/10 = 0.7$ but claiming that this is also the probability of a male baby is now not justified because the number of babies is too small.

Subjective probability: The classical and relative frequency definitions of probabilities cannot be used in *all* situations where probability statements are required; for example, the question, 'what is the probability that life exists on a planet circling the nearest star, *Proxima Centauri?*' The classical definition cannot provide an answer and there are no relative frequency data available that might help. In such circumstances we can only express a subjective opinion about the required probability, i.e., assign a subjective probability. We might use experience, intuition or a hunch to arrive at a value but the assigned probability will be subjective and different people can be expected to assign different probabilities to the same event. Subjective probabilities express a person's degree of belief in the event and are often referred to as *personal probabilities*.

Addition rule for probabilities: Often the events for which we require probabilities are made up of a number of simpler events and so we need to consider how to combine the probabilities of these simpler events to arrive at the probability of the more complex event. For *mutually exclusive events* (events that cannot occur together), $E_1, E_2, \ldots, E_r$, we have the following addition rule:

$$\Pr(E_1 \text{ or } E_2 \text{ or } \ldots \text{ or } E_r) = \Pr(E_1) + \Pr(E_2) + \ldots \Pr(E_r)$$

EXAMPLE 2.3

What is the probability of rolling a fair die (one in which all 6 faces are equally likely) and getting an even number?

SOLUTION

Pr(2 or 4 or 6) = Pr(2)+Pr(4) + Pr(6) = 1/6 + 1/6 + 1/6 = 1/2

(NB: When events are not mutually exclusive the addition rule no longer holds; for example, for two events E_1 and E_2 that *can* occur together the probability that either E_1 or E_2 occurs is given by

$$\Pr(E_1 \text{ or } E_2) = \Pr(E_1) + \Pr(E_2) - \Pr(E_1 \text{ and } E_2).)$$

Multiplication rule for probabilities: Now consider events that can happen together but which are *independent*, i.e., the occurrence or otherwise of any event does not affect the occurrence or otherwise of any other event. Here the probability that all events $E_1, E_2,\ldots,E_r$ occur simultaneously is found from the following multiplication rule:

$$\Pr(E_1 \text{ and } E_2 \text{ and } \ldots\text{and } E_r) = \Pr(E_1)\Pr(E_2)\ldots\Pr(E_r)$$

EXAMPLE 2.4

What is the probability of getting a triple 6 when rolling three fair dice?

SOLUTION

Pr(triple 6) = Pr(6 with die 1)Pr(6 with die 2)Pr(6 with die 3) = 1/6 × 1/6 × 1/6 = 1/216

2.2 ODDS AND ODDS RATIOS

Odds and probability: Gamblers (and some statisticians) prefer to quantify their uncertainty about an event in terms of *odds* rather than probabilities although the two are actually completely synonymous. The exact relationship between odds and probability is as follows:

- An event with odds of 'F to 1 in favour (odds-on)' has probability $F/(F + 1)$,
- An event with odds of 'A to 1 against' has probability $1/(A + 1)$,
- An event with probability P implies that the odds in favour are $P/(1 - P)$ to 1, whereas the odds against are $1/P - 1$.

EXAMPLE 2.5

What are the odds for an event with probability 1/5?

SOLUTION

The odds against are $1/0.2 - 1 = 4$ to 1 against. This simply expresses the fact that the probability of the event not occurring is four times that of it occurring.

EXAMPLE 2.6

What are the odds for an event with probability 4/5?

SOLUTION

The odds in favour are 0.8/0.2, i.e., 4 to 1; here the probability of the event occurring is four times that of it not occurring.

2.3 PERMUTATIONS AND COMBINATIONS

Permutations: If r objects are selected from a set of n total objects (and not replaced after each object is selected) any particular ordered arrangement of the r objects is called a permutation. The total number of possible permutations of r objects selected from n objects is denoted $^n p_r$ and given by

$$^n p_r = \frac{n!}{(n - r)!}$$

where $n!$ (read n factorial) is given by $n! = n(n-1)...3 \times 2 \times 1$ with $0!$ defined as one.

The number of permutations of all the objects taken together, ^np_n, is $n!$

Combinations: If r objects are selected from n total objects, and if the order of the r objects is not important, then each possible selection of r objects is called a combination, denoted by ^nc_r and given by

$$^nc_r = \frac{n!}{r!(n-r)!}$$

(^nc_r is often called a *binomial coefficient*.)

The numbers of permutations and combinations of r objects in a total n objects are often useful when assigning probabilities to events.

EXAMPLE 2.7

The four letters s, e, n and t are placed in a row at random. What is the chance of their standing in such order to form an English word?

SOLUTION

The four letters can be arranged in $4! = 24$ different permutations, all of which are equally likely. Only four of these arrangements, *sent*, *nest*, *nets*, *tens*, produce an English word so the required probability is $4/24 = 1/6$.

EXAMPLE 2.8

A bag contains five white and four black balls. If balls are drawn out of the bag one by one, what is the probability that the first will be white, the second black and so on alternately?

SOLUTION

Assuming white balls are all alike and black balls are all alike, the number of possible combinations of ways in which the nine balls can be drawn from the bag is $9!/(4! \times 5!) = 126$. The balls are equally likely to be drawn in any of these ways; only 1 of the 126 possibilities is the alternate order required, so the probability of white, followed by black, followed by white, etc. is $1/126$.

EXAMPLE 2.9

Four cards are drawn from a pack of 52 cards. What is the probability that there will be one card of each suit?

SOLUTION

Four cards can be selected from the pack in $52!/(4! \times 48!) = 270{,}725$ ways. But four cards can be selected to be one of each suit in only $13 \times 13 \times 13 \times 13 = 28{,}561$ ways. So the required probability is $28{,}651/270{,}725$, which is just a little over $1/10$.

2.4 CONDITIONAL PROBABILITIES AND BAYES' THEOREM

Conditional probability: The probability that an event A occurs given the outcome of some other event B on which A is dependent; usually written as $\Pr(A|B)$. It is not, of course, necessary for $\Pr(A|B)$ to be equal to $\Pr(B|A)$. For example, $\Pr(\text{spots}|\text{suffering from measles})$ is likely to be very different from $\Pr(\text{suffering from measles}|\text{spots})$. If $\Pr(A|B) = \Pr(A)$ then the events A and B are independent. The probability that both of two dependent events A and B occur is

$$\Pr(A \text{ and } B) = \Pr(A) \times \Pr(B|A)$$

EXAMPLE 2.10

When rolling two dice what is the probability that the number on the die rolled second is larger than the number on the die rolled first?

SOLUTION

FIRST DIE SCORE	SECOND DIE HIGHER IF RESULT IS	PROBABILITY THAT SECOND DIE HIGHER
1	2 or 3 or 4 or 5 or 6	$1/6 + 1/6 + 1/6 + 1/6 + 1/6$
2	3 or 4 or 5 or 6	$1/6 + 1/6 + 1/6 + 1/6$
3	4 or 5 or 6	$1/6 + 1/6 + 1/6$
4	5 or 6	$1/6 + 1/6$
5	6	$1/6$
6	Cannot have a higher score	0

$\Pr(\text{first die shows a 1 and second die is higher}) = 1/6 \times 5/6 = 5/26$. Now find corresponding probabilities for different first die scores and then add to give the result as $15/36$.

Bayes's theorem: A theorem relating conditional probabilities:

$$\Pr(A|B) = \frac{\Pr(B|A)\Pr(A)}{\Pr(B)}$$

EXAMPLE 2.11

A cab was involved in a hit-and-run accident at night. Two cab companies, the Green and the Blue, operate in the city. The following facts are known:

- 85% of the cabs in the city are Green and 15% are Blue.
- A witness identified the cab as Blue. The court tested the reliability of the witness under the same circumstances that existed on the night of the accident and concluded that the witness correctly identified each one of the two colours 80% of the time and failed 20% of the time.

What is the probability that the cab involved in the accident was actually Blue?

SOLUTION

Letting A be the event that a cab is Blue and B the event that the witness says he sees a Blue cab then, equating percentages to probabilities, what we know is

$$\Pr(A) = 0.15, \Pr(B|A) = 0.80.$$

The unconditional probability of the witness saying that he saw a Blue cab is

$$\Pr(B) = 0.15 \times 0.80 + 0.85 \times 0.20 = 0.29$$

Applying Bayes' theorem we can now find what we require, namely, the probability that the cab is Blue given that the witness says it is Blue, $\Pr(A|B)$,

$$\Pr(A|B) = \frac{\Pr(B|A)\Pr(A)}{\Pr(B)} = \frac{0.80 \times 0.15}{0.29} = 0.41$$

So the probability that the cab is actually Blue given that the witness identifies the cab as Blue is less than a half. Despite the evidence the eyewitness offers, the hit-and-run cab is more likely to be Green than Blue! The report of the eyewitness has, however, increased the probability that the offending cab is Blue from its value of 0.15 in the absence of any evidence (known as the *prior probability*) to 0.41 (known as the *posterior probability*).

Conditional probabilities are of particular importance in medical diagnoses, i.e., using medical tests to diagnose the condition of a patient. The sensitivity and specificity of such a test are defined as

Sensitivity = Pr(Test positive|Patient has the disease)
Specificity = Pr(Test negative|Patient does not have the disease)

2.5 RANDOM VARIABLES, PROBABILITY DISTRIBUTIONS AND PROBABILITY DENSITY FUNCTIONS

2.5.1 Random Variable

A rule that assigns a unique number to each outcome of an experiment; these numbers are called the *values* of the random variable. The value of a random variable will vary from trial to trial as the experiment is repeated. Random variables can be either *discrete,* meaning that they can only take specific numerical values, or *continuous*, meaning that they can take any value within a continuous range or interval. An example of a discrete random variable is the number of heads obtained when a coin is tossed 10 times; here the random variable can take values 0,1,2,3,...,10. An example of a continuous random variable is age at death in humans. Random variables are usually denoted by capital letters, for example, X, with particular outcomes being represented by the corresponding lower case letter, here, x.

2.5.2 Probability Distribution

If X represents a discrete random variable which can take on m values, $x_1, x_2,..., x_m$, the probability distribution (sometimes probability function) of X is a function f where $f(x_i) = \text{Pr}(X = x_i)$, $i = 1,2...m$, $0 \le f(x) \le 1$ and $\sum_{i=1}^{m} f(x_i) = 1$.

(NB: f is used below to denote a variety of *different* probability functions.)

EXAMPLE 2.12

What is the probability distribution for the number of heads obtained in n independent tosses of a fair coin?

SOLUTION

The random variable, X, here takes the values $0,1,2,...,n$. The probability of a head on any toss is ½ as is the probability of a tail. The probability of obtaining r heads in the n tosses is found by multiplying the probability of r heads and $n - r$ tails, namely, $(0.5)^r \times (0.5)^{(n-r)}$ by the number of combinations of r heads in n tosses, ^nc_r, to give the required probability function as

$$\Pr(X = r) = f(r) = {}^nc_r \times (0.5)^r \times (0.5)^{(n-r)}$$

$$= \frac{n!}{r!(n-r)!}(0.5)^n, \ r = 0,1,2,...n$$

Binomial distribution: The result in the example above can be generalized to a situation where there are n independent trials with the probability of a success on each trial being p, to give what is known as the binomial distribution:

$$f(r) = \frac{n!}{r!(n-r)!} p^r (1-p)^{(n-r)}, r = 0,1,2,...n$$

Poisson distribution: Another important probability distribution for a discrete random variable, X, which arises as a limiting form of the binomial distribution as $n \to \infty$ and $p \to 0$ with $\lambda = np$ is the Poisson distribution:

$$\Pr(X = x) = f(x) = \frac{\lambda^x e^{-\lambda}}{x!}, x = 0,1,2...,\infty$$

The Poisson distribution is also important in its own right for assigning probabilities to rare events of some sort occurring randomly in time or space.

EXAMPLE 2.13

How do the probabilities given by the Poisson distribution compare with those from a binomial distribution with $n = 60$ and $p = 0.05$?

SOLUTION

NUMBER OF EVENTS	BINOMIAL DISTRIBUTION WITH N = 60 AND p = 0.05	POISSON DISTRIBUTION WITH λ = 3
0	0.046	0.050
1	0.145	0.149
2	0.226	0.224
3	0.230	0.224
4	0.172	0.168
5	0.102	0.101
6	0.049	0.050
7	0.020	0.022
8	0.007	0.008
9	0.002	0.003

Parameters: The terms p in the binomial distribution and λ in the Poisson distribution are *parameters* of the distributions where a parameter is a quantity that characterizes the associated random variable.

(Other important probability distributions for discrete random variables are given in the Summary.)

2.5.3 Probability Density Function

For a continuous random variable, X, there is an associated probability density function (pdf), $f(x)$, that allows the probability of the variable lying between two values to be calculated, but the probability of X taking any particular value is zero. The probability density function is such that

$$\Pr(a < X < b) = \int_{a}^{b} f(x)dx \text{ with } f(x) \geq 0 \text{ and } \int_{l}^{u} f(x)dx = 1$$

where l and u are, respectively, the lower and upper values that X can take.

Uniform (rectangular) probability density function: A very simple pdf is the uniform or rectangular which is the pdf for a continuous random variable that can take any value in an interval $[a,b]$ and for which the probability of assuming a value in a subinterval of $[a,b]$ is the same for all subintervals of equal length. The uniform pdf is given by

$$f(x) = \frac{1}{b-a} \text{ for } a \le x \le b \text{ and } f(x) = 0 \text{ for } x < a, x > b; a < b$$

A graph of the uniform pdf is shown in Figure 2.1.

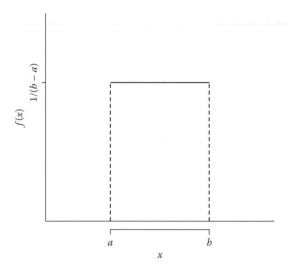

FIGURE 2.1 Uniform density function.

EXAMPLE 2.14

What is the probability that a random variable with a uniform pdf in the interval [2,5] takes a value between 3 and 4?

SOLUTION

The pdf of the random variable is $f(x) = 1/3$; the required probability is given by

$$\Pr(3 < x < 4) = \int_{3}^{4} \frac{1}{3} dx = \frac{1}{3}[x]_{3}^{4} = \frac{1}{3}$$

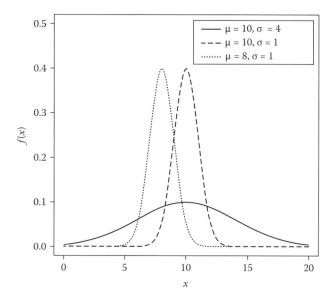

FIGURE 2.2   Some normal density functions.

Normal (Gaussian) pdf: A very important pdf in statistics is the normal density function (also often called the Gaussian). The density function is given by

$$f(x) = \frac{1}{\sigma\sqrt{2\pi}} \exp\left[-\frac{1}{2} \frac{(x-\mu)^2}{\sigma^2} \right], \quad -\infty < x < \infty$$

The terms μ and σ are parameters of the normal density function and the 'bell-shape' of the density changes as the parameters take different values, as shown in Figure 2.2. [NB: The normal density with parameters μ and σ is often written as $N(\mu, \sigma)$ or $N(\mu, \sigma^2)$.]

Standard normal: A normal density function with $\mu = 0$ and $\sigma = 1$ is called the standard normal density. Areas under *any* normal density, which are required to give probabilities that the associated random variable falls in some interval, can be calculated from the standard normal.

EXAMPLE 2.15

A random variable, X, has a normal density function with $\mu = 10$ and $\sigma = 2$. What is the probability that X takes a value between 8 and 12?

SOLUTION

A normal density with parameters μ and σ can be transformed into a standard normal density by setting $z = \dfrac{x - \mu}{\sigma}$. So the area between 8 and 12 under the N(10,2) density is equivalent to the area under a standard normal density between the limits $(8 - 10)/2 = -1$ and $(12 - 10)/2 = 1$. Areas under the standard normal density can be found from tables or in statistical software packages. In this case the area, and hence the required probability, is 0.68.

(Other important pdfs for continuous random variables are given in the Summary.)

Cumulative distribution function: The cumulative distribution function (cdf) of a random variable X gives the probability that X takes on some value less than or equal to y. For a discrete random variable with probability distribution function, $f(x)$, the cdf is given by

$$\Pr(X \le y) = \sum_{x \le y} f(x)$$

and for a continuous random variable the cdf is given by

$$\Pr(X < y) = \int_{l}^{y} f(x)dx$$

(NB: The term *probability distribution* is often used instead of probability density function when discussing continuous random variables and although it is perhaps preferable to distinguish the two, probability distribution will often be used in this way in this book.)

2.6 EXPECTED VALUE AND MOMENTS

Moments: Quantities used to summarize and quantify various features of a probability distribution or density function. The rth moment of a random variable, X, about zero is simply the average value or *expected value* of X^r; for a discrete random variable taking values say from zero to n, this expected value is given by

$$E(X^r) = \sum_{k=0}^{n} k^r \Pr(X = k)$$

For a continuous random variable with pdf, $f(x)$, and taking values in the range $[l,u]$ the rth moment about zero is given as

$$E(X^r) = \int_{l}^{u} x^r f(x) dx$$

The first moment about zero ($r = 1$) is the *mean* of the distribution of possible values for X; it is also called the expected value of X.

EXAMPLE 2.16

What is the expected value of a random variable with a binomial distribution?

SOLUTION

The required expected value is given by

$$E(X) = \sum_{k=0}^{n} k \frac{n!}{k!(n-k)!} p^k (1-p)^{n-k}$$

$$= np\{(1-p)^{n-1} + (n-1)p(1-p)^{n-2} + \ldots p^{n-1}\}$$

$$= np[(1-p) + p]^{n-1}$$

$$= np$$

EXAMPLE 2.17

What is the expected value of a random variable with a Poisson distribution?

SOLUTION

The required expected value is given by

$$E(X) = \sum_{k=0}^{\infty} k \frac{\lambda^k e^{-\lambda}}{k!} = \lambda e^{-\lambda} \left\{ 1 + \lambda + \frac{\lambda^2}{2!} + \frac{\lambda^3}{3!} + \ldots \right\} = \lambda e^{-\lambda} e^{\lambda} = \lambda$$

EXAMPLE 2.18

What is the expected value of a random variable with a uniform density function in the interval [a,b]?

SOLUTION

The required expected value is given by

$$E(X) = \int_a^b x \frac{1}{b-a} dx = \frac{1}{b-a} \left[\frac{x^2}{2} \right]_a^b = \frac{1}{2(b-a)}(b^2 - a^2) = \frac{1}{2}(b+a)$$

In the same way the expected value of a random variable having a normal pdf is the parameter μ.

Central moments: If we now use the term μ for the expected value of any random variable, then we can define the rth central moment, μ_r, of a discrete random variable taking values $0,1,2...n$ as

$$\mu_r = \sum_{k=0}^{n} (k - \mu)^r \Pr(X = k)$$

and for a continuous random variable taking values in the interval $[l,u]$ and with density function $f(x)$ as

$$\mu_r = \int_l^u (x - \mu)^r f(x) dx$$

Variance: The first central moment is zero by definition. The second central moment, $E(X - \mu)^2$, is called the variance of the random variable and its positive square root is known as the *standard deviation* of X. The standard deviation provides a measure of the level of dispersion in the distribution of possible values of X. The ratio of the standard deviation to the mean (often multiplied by 100) is known as the *coefficient of variation*. (NB: The variance can also be written as $E(X^2) - [E(X)]^2$.)

EXAMPLE 2.19

What is the variance of a random variable having a binomial distribution?

SOLUTION

As we already know the mean to be np we just need to find $E(X^2)$; this is given as

$$E(X^2) = \sum_{k=0}^{n} k^2 \frac{n!}{k!(n-k)!} p^k (1-p)^{n-k} = np + n(n-1)p^2$$

So the variance of the random variable is

$$\mu_2 = np + n(n-1)p^2 - (np)^2 = np(1-p)$$

EXAMPLE 2.20

What is the variance of a random variable having a Poisson distribution?

SOLUTION

We first need to find $E(X^2)$:

$$E(X^2) = \sum_{k=0}^{\infty} k^2 \frac{\lambda^k e^{-\lambda}}{k!} = \lambda + \lambda^2$$

So the variance of the random variable is $\lambda + \lambda^2 - \lambda^2 = \lambda$. The mean and variance of a Poisson random variable are equal.

EXAMPLE 2.21

What is the variance of a random variable having a uniform density in the interval $[a,b]$?

SOLUTION

First we must find $E(X^2)$ which is given by

$$E(X^2) = \int_a^b x^2 \frac{1}{b-a} dx$$

$$= \frac{1}{b-a} \left[\frac{1}{3} x^3 \right]_a^b$$

$$= \frac{1}{3(b-a)} [(b-a)(b^2 + ab + a^2)]$$

$$= \frac{(b^2 + ab + a^2)}{3}$$

Therefore the variance is given by

$$\mu_2 = \frac{(b^2 + ab + a^2)}{3} - \left(\frac{a+b}{2} \right)^2 = \frac{(b-a)^2}{12}$$

The variance of a normal distribution is the square of its parameter, σ.
Relating central moments and moments about zero: Central moments can be obtained from moments about zero by the formula

$$\mu_r = \sum_{j=0}^r (-1)^j \, {}^r c_j \mu^j E(X^{r-j})$$

For example,

$$\mu_2 = E(X^2) - \mu^2$$

$$\mu_3 = E(X^3) - 3\mu E(X^2) + 2\mu^3$$

$$\mu_4 = E(X^4) - 4\mu E(X^3) + 6\mu^2 E(X^2) - 3\mu^4$$

These formulae can also be inverted to give the moments about the origin in terms of the central moments

$$E(X^2) = \mu_2 + \mu^2$$

$$E(X^3) = \mu_3 + 3\mu_2\mu + \mu^3$$

$$E(X^4) = \mu_4 + 4\mu_3\mu + 6\mu_2\mu^2 + \mu^4$$

Shape of probability distributions: Particular functions of moments are also used to characterize the shapes of probability distributions and pdfs by way of their *skewness* (lack of symmetry) and their *kurtosis* (the extent to which the peak of a unimodal pdf departs from the shape of a normal pdf.) A measure of skewness is $\mu_3/\mu_2^{3/2}$ and a measure of kurtosis is μ_4/μ_2^2.

Rules for finding expected values and variances of sums of random variables:

$E(aX) = aE(X)$ where a is a scalar

$$E(X_1 + X_2 + \ldots + X_m) = E(X_1) + E(X_2) + \ldots + E(X_m)$$

$\text{Var}(aX) = a^2\text{Var}(X)$ where a is a scalar

$\text{Var}(X_1 + X_2 + \ldots + X_m) = \text{Var}(X_1) + \text{Var}(X_2) \ldots + \text{Var}(X_m)$ if the random variables X_1 to X_m are mutually independent. When the variables are not independent the formula becomes

$$\text{Var}\left(\sum_{i=1}^{m} X_i\right) = \sum_{i=1}^{m} \text{Var}(X_i) + \sum_i \sum_j \text{Cov}(X_i, X_j)$$

where

$\text{Cov}(X_i, X_j) = E(X_i - \mu_i)(X_j - \mu_j)$. This is known as the *covariance* of X_i and X_j (for further details see Chapter 10).

2.7 MOMENT-GENERATING FUNCTION

Moment-generating function: The moment-generating function of a random variable, X, $M_X(t)$, is defined as

$$M_X(t) = E[\exp(tX)]$$

If $M_X^{(r)}(t)$ is used to denote the rth derivative of $M_X(t)$ with respect to t, then the rth moment of X about the origin is given by

$$E(X^r) = M_X^{(r)}(0) \text{ for } r = 1,2,\ldots,$$

EXAMPLE 2.22

What is the moment-generating function of a random variable having the binomial distribution?

SOLUTION

The moment-generating function in this case is

$$M_X(t) = E[\exp(tX)]$$

$$= \sum_{k=0}^{n} e^{kt} \frac{n!}{k!(n-k)!} p^k (1-p)^{n-k}$$

$$= [(1-p) + pe^t]^n$$

The mean of the binomial can be found from $M_X^{(1)}(t)$, which is $M_X^{(1)}(t) = n[(1-p) + pe^t]^{n-1} pe^t$ and evaluating this expression at $t = 0$ gives the mean as np.

EXAMPLE 2.23

What is the moment-generating function of a Poisson distribution?

SOLUTION

The moment-generating function here is

$$M_X(t) = \sum_{k=0}^{\infty} e^{kt} \frac{\lambda^k e^{-\lambda}}{k!} = \exp[\lambda(e^t - 1)]$$

EXAMPLE 2.24

What is the moment-generating function of a random variable with a normal distribution?

SOLUTION

The required moment-generating function is

$$M_X(t) = \int_{-\infty}^{\infty} e^{xt} N(\mu,\sigma)dx = \exp\left[\mu t + \frac{\sigma^2 t^2}{2}\right]$$

The moment-generating function uniquely determines a probability distribution or pdf so that if random variables X and Y have, respectively, moment-generating functions, $M_X(t)$ and $M_Y(t)$, then the two variables have the same probability distribution if and only if $M_X(t) = M_Y(t)$. And if $X_1, X_2, \ldots, X_n$ are independent random variables then, for example, $M_{X_i - X_j}(t) = M_{X_i}(t) M_{X_j}(-t)$ and $M_{\sum_{i=1}^{n} X_i}(t) = M_{X_1}(t) M_{X_2}(t) \ldots M_{X_n}(t)$.

EXAMPLE 2.25

What is the moment-generating function of the sum of two independent Poisson random variables?

SOLUTION

The first variable, X_1, has a Poisson distribution with parameter λ_1 and the second variable, X_2, has a Poisson distribution with parameter λ_2. The moment-generating function of $X_1 + X_2$ is the product of the two moment-generating functions, i.e.,

$$e^{\lambda_1(e^t - 1)} e^{\lambda_2(e^t - 1)} = e^{(\lambda_1 + \lambda_2)(e^t - 1)}$$

This is the form of the moment-generating function of a Poisson random variable with parameter equal to $\lambda_1 + \lambda_2$. We can therefore infer that the sum of two independent Poisson variables has a Poisson distribution with a parameter equal to the sum of the individual distribution parameters.

(NB: The *characteristic function* of a random variable defined as $\phi_X(t) = E[\exp(itX)]$ where $i = \sqrt{-1}$ has many of the same uses as the moment-generating function but is often preferable in theoretical work because it exists for *any* probability distribution.)

2.8 SUMMARY

Relative frequency definition of probability: $p = \lim\limits_{n \to \infty} \dfrac{s}{n}$

Probability addition rule for mutually exclusive events: $E_1, E_2, \ldots, E_m$: $\Pr(E_1 \text{ or } E_2 \text{ or } \ldots E_m) = \Pr(E_1) + \Pr(E_2) + \ldots + \Pr(E_m)$

Probability multiplication rule for independent events: $E_1, E_2, \ldots, E_m$: $\Pr(E_1 \text{ and } E_2 \text{ and } \ldots E_m) = \Pr(E_1) \times \Pr(E_2) \times \ldots \times \Pr(E_m)$

Number of permutations of r objects from n objects: $^nP_r = \dfrac{n!}{(n-r)!}$

Number of combinations of r objects from n objects: $^nC_r = \dfrac{n!}{r!(n-r)!}$

Bayes' theorem: $\Pr(A|B) = \Pr(B|A)\,\Pr(A)/\Pr(B)$

Binomial distribution: $\Pr(X = x) = \dfrac{n!}{x!(n-x)!}\,p^x(1-p)^{n-x}$

$E(X) = np$, $\mu_2 = np(1-p)$, $M_X(t) = \left[(1-p) + pe^t\right]^n$

Poisson distribution: $\Pr(X = x) = \dfrac{\lambda^x e^{-\lambda}}{x!}$

$E(X) = \lambda$, $\mu_2 = \lambda$, $M_X(t) = \exp[\lambda(e^t - 1)]$

Geometric distribution: Distribution of the number of 'successes' in a series of independent trials until the first failure and where the probability of a success on any trial is p: $\Pr(X = x) = (1-p)p^x, x = 0, 1, \ldots$

$E(X) = \dfrac{p}{1-p}$, $\mu_2 = \dfrac{p}{(1-p)^2}$, $M_X(t) = \dfrac{1-p}{1-pe^t}$

Negative binomial distribution: Distribution of the number of independent trials to obtain m 'successes' where the probability of a success on any trial is p: $\Pr(X = m+k) = {}^{m+k-1}C_{m-1}\,p^m(1-p)^k, k = 0, 1, 2, \ldots$

$$E(X) = \frac{m(1-p)}{p}, \ \mu_2 = \frac{m(1-p)}{p^2}, \ M_X(t) = \left[\frac{1-(1-p)e^t}{p}\right]^{-m}$$

Hypergeometric distribution: A probability distribution associated with sampling without replacement from a population of finite size. If the population consists of r elements of one kind and $N - r$ of another and the random variable X is the number of elements of the first kind obtained when a random sample of n is drawn:
$\Pr(X = k) = {}^r c_k \ {}^{N-r} c_{n-k} \ / \ {}^N c_n$, for $\max(0, n - N + r) \le k \le \min(n, r)$

$$E(X) = \frac{nr}{N}, \ \mu_2 = \frac{nr(1 - r/n)(N - n)}{N(N-1)}$$

Uniform distribution in interval *[a,b]*: $f(x) = \dfrac{1}{b-a}$

$$E(X) = \frac{a+b}{2}, \ \mu_2 = \frac{(b-a)^2}{12}, \ M_X(t) = \frac{e^{tb} - e^{ta}}{(b-a)t}$$

Normal distribution: $f(x) = \dfrac{1}{\sigma\sqrt{2\pi}} e^{-\frac{1}{2}\left(\frac{x-\mu}{\sigma}\right)^2}, \ -\infty < x < \infty$

$$E(X) = \mu, \ \mu_2 = \sigma^2, \ M_X(t) = \exp\left(\mu t + \frac{\sigma^2 t}{2}\right)$$

Beta distribution: $f(x) = \dfrac{\Gamma(\alpha+\beta)}{\Gamma(\alpha)\Gamma(\beta)} x^{\alpha-1}(1-x)^{\beta-1}, \ 0 < x < 1$

$$E(X) = \frac{\alpha}{\alpha+\beta}, \ \mu_2 = \frac{\alpha\beta}{(\alpha+\beta)(\alpha+\beta+1)}$$

Exponential distribution: $f(x) = \lambda e^{-\lambda x}, \ x > 0$

$$E(X) = \frac{1}{\lambda}, \ \mu_2 = \frac{1}{\lambda^2}, \ M_X(t) = \frac{1}{1 - t/\lambda}$$

Gamma distribution: $f(x) = \dfrac{1}{\Gamma(\alpha)} x^{\alpha-1} e^{-x}$, $x > 0$

$E(X) = \alpha$, $\mu_2 = \alpha$, $M_X(t) = \dfrac{1}{(1-t)^\alpha}$

Mean: $E(X)$

Variance: $\mu_2 = E(X^2) - [E(X)]^2$

Moment-generating function: $M_X(t) = E[\exp(tX)]$

Moment-generating function of the sum of independent random variables is the product of the moment-generating functions of each variable: $M_{X_1+X_2+\ldots+X_m}(t) = M_{X_1}(t) M_{X_2}(t) \ldots M_{X_m}(t)$

SUGGESTED READING

Andrei, J (2001) *Mathematics of Chance*, Wiley, New York.

Balakrishnan, N and Nevzorov,VB (2003) *A Primer of Statistical Distributions*, Wiley, New York.

Everitt, BS (2008) *Chance Rules*, 2nd edition, Springer, New York.

Larsen, RJ and Marx, ML (2011) *Introduction to Mathematical Statistics and Its Applications*, Prentice Hall, Upper Saddle River, NJ.

Estimation 3

3.1 POINT ESTIMATION

Point estimation: The process of using values of a random variable sampled from a population in which the random variable is assumed to follow a particular probability distribution function, to estimate the unknown value of a parameter of the probability distribution function.

A *point estimate* is a single number.

The essential features of a point estimation problem are

- Some sample values of a random variable, i.e., some data.
- An assumed population probability distribution for the random variable.
- The population probability distribution is characterised by one or more unknown parameter values.
- An estimating formula or *estimator* for the unknown parameter.
- The value of the estimator given the data, i.e., the *estimate* of the parameter.

(NB: Note the essential difference here—the estimate is a number obtained from the data while the estimator is a random variable expressing an estimating formula; *see* Section 3.2)

An example of two intuitively sensible point estimators is the sample mean and sample variance for the parameters μ and σ^2 of an assumed normal probability density function for the population from which the sample is drawn.

As a second example consider the data below giving the number of women in each of 100 queues of length 10 observed at a London underground station.

Count	0	1	2	3	4	5	6	7	8	9	10
Frequency	1	3	4	23	25	19	18	5	1	1	0

A reasonable assumption for the probability distribution of the number of women in a queue of length 10 is a binomial with $n = 10$ and unknown parameter p for the probability of a woman passenger using the London underground transport system during the time that the observations were made. An intuitive estimator for the parameter is the number of females divided by the total number of passengers, i.e.,

$$\hat{p} = \frac{\sum_{i=1}^{10} f_i c_i}{10 \times 100}$$

where f_i and c_i represent, respectively, the frequencies and the counts. Performing the calculation gives $\hat{p} = 0.435$; note the 'hat' nomenclature for representing the estimator of a parameter to distinguish it from the parameter itself.

(NB: One of the assumptions underlying the assumption of a binomial distribution here is that the observations are independent, i.e., the gender of any person in a queue does not affect, and is not affected by, the gender of the others. With queues this may not be realistic because of the possibility of male–female couples queuing together.)

3.2 SAMPLING DISTRIBUTION OF THE MEAN AND THE CENTRAL LIMIT THEOREM

Distribution of the sample mean of samples from a normal distribution: When sampling values of a random variable from a population where the probability density function of the random variable has expected value μ and variance σ^2, the mean of the sample values is used to estimate μ. The mean of n sample values is itself a random variable and we can ask what are its expected value and its variance?

$$E(\bar{x}) = \frac{E(x_1) + E(x_2) + \ldots + E(x_n)}{n} = \mu$$

$$\mathrm{Var}(\bar{x}) = \frac{1}{n^2}[\mathrm{Var}(x_1) + \mathrm{Var}(x_2) + \ldots + \mathrm{Var}(x_n)] = \frac{\sigma^2}{n}$$

If we assume further that the distribution of the random variable in the population is normal, then we have the result that the distribution of the mean of n sample values is N(μ, σ^2/n).

Sampling distribution: The probability distribution of a statistic, for example, the sample mean. Figure 3.1 shows the sampling distribution of the sample mean for different sample sizes from N(6,4). As sample size increases, the sampling distribution of the sample mean becomes more and more 'pinched'; the sample

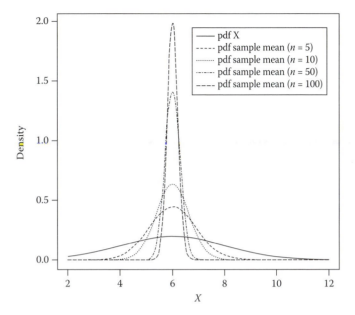

FIGURE 3.1 Sampling distributions of the sample mean for different values of *n*: samples from N(6,4).

mean provides a more precise estimate of the population mean as sample size increases.

Central limit theorem: If a random variable X has population mean μ and population variance σ^2, the mean of a sample of n values from the population is approximately normally distributed with expected value μ and variance σ^2/μ for sufficiently large n. (NB: The probability distribution of X is *not* confined to be a normal distribution.)

EXAMPLE 3.1

How large does n have to be for the mean of samples of size n from a uniform distribution in the interval [0,1] to have an approximate normal distribution? The population distribution here is clearly non-normal.

SOLUTION

We can investigate this by repeated sampling from the uniform density with a different value of n. Some results are shown in Figure 3.2. It appears that with a sample size of 10 and larger the histogram of the sample mean approximates to the normal shape even in this case where the population distribution is very non-normal.

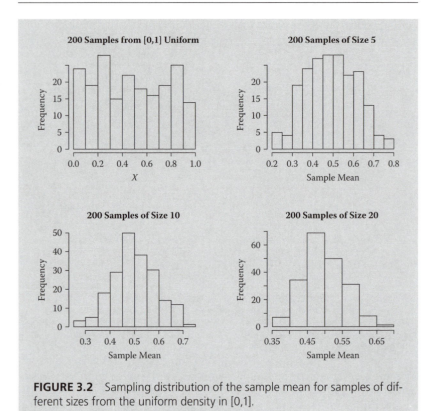

FIGURE 3.2   Sampling distribution of the sample mean for samples of different sizes from the uniform density in [0,1].

(NB: The central limit theorem provides some reassurance about the normality of the sampling distribution of the sample mean for reasonably sized samples, even when the underlying distribution is not normal. This is important in *inference* and *interval estimation, see* later.)

3.3 ESTIMATION BY THE METHOD OF MOMENTS

Method of moments: The basis of estimation by the method of moments is the matching of corresponding sample and population moments. For example, for a random variable with population distribution $N(\mu, \sigma^2)$, the method of moments matches the unknown parameters μ and σ^2 precisely to their sample analogues $\bar{x}$ and s^2 to give

$$\hat{\mu} = \bar{x} \text{ and } \hat{\sigma}^2 = s^2$$

EXAMPLE 3.2

What is the method of moments estimator for the parameter p of a geometric distribution?

SOLUTION

The expected value for a geometric probability distribution is $1/p$; matching the first population moment with the first sample moment we obtain

$$\bar{x} = \frac{1}{\hat{p}}$$

leading to the following estimator for p,

$$\hat{p} = \frac{1}{\bar{x}}$$

It is not always obvious which moments we match; for example, the variance of a geometric distribution is $(1-p)/p^2$ and so we could match the sample variance to this value,

$$s^2 = \frac{1 - \hat{p}}{\hat{p}^2}$$

which can be solved to give

$$\hat{p} = \frac{\sqrt{1 + 4s^2} - 1}{2s^2}$$

This is also a method of moments estimator for p but for any particular sample will not be equal to the first estimator, $\hat{p} = 1/\bar{x}$. Which estimator is preferable will depend on properties of their sampling distribution, in particular the variance of this distribution, with the estimator having a smaller variance being preferred.

The method of moments approach to estimation is intuitive and straightforward but there are times when it fails or cannot be applied at all. The method has been largely displaced by methods known to be more statistically efficient, primarily *maximum likelihood*.

3.4 ESTIMATION BY MAXIMUM LIKELIHOOD

Maximum likelihood estimation (MLE): A method of estimation based on trying to find from the sample values a numerical value of an unknown parameter that is 'most likely' to have produced the data, where 'most likely' is judged from the *likelihood function*.

Likelihood: The probability of the observed data given the probability distribution of the population from which the sample is taken. For a sample $x_1, x_2, \ldots, x_n$ drawn independently from a population with probability distribution $f(x: \theta)$, with an unknown parameter θ, the likelihood, L, is given by

$$L(x_1, x_2, \ldots, x_n; \theta) = f(x_1; \theta) f(x_2; \theta) \ldots f(x_n; \theta) = \prod_{i=1}^{n} f(x_i; \theta)$$

The value $\hat{\theta}$ of θ at which the likelihood is maximised is known as the *maximum likelihood estimator* of θ. In practice it is often more convenient to maximize the logarithm of the likelihood, the *log-likelihood*.

EXAMPLE 3.3

What is the maximum likelihood estimator of the parameter p of a geometric probability distribution, given n observations, $x_1, x_2, \ldots x_n$?

SOLUTION

The likelihood in this case is

$$L = \prod_{i=1}^{n} p(1-p)^{x_i - 1} = p^n \prod_{i=1}^{n} (1-p)^{x_i - 1}$$

The log-likelihood, l, is given by

$$l = n \log(p) + \sum_{i=1}^{n} (x_i - 1) \log(1-p)$$

$$= n \log(p) + \sum_{i=1}^{n} x_i \log(1-p) - n \log(1-p)$$

To find the value of p that maximises l we need to solve the equation resulting from setting $\dfrac{dl}{dp} = 0$:

$$\frac{dl}{dp} = \frac{n}{p} - \frac{\sum_{i=1}^{n} x_i}{(1-p)} + \frac{n}{(1-p)}$$

Solving $\dfrac{dl}{dp} = 0$ in this case gives the estimator of p as

$$\hat{p} = \frac{n}{\sum_{i=1}^{n} x_i} = \frac{1}{\overline{x}}$$

EXAMPLE 3.4

What is the maximum likelihood estimator of the parameter of an exponential density function?

SOLUTION

The likelihood here is given by

$$L = \prod_{i=1}^{n} \lambda e^{-\lambda x_i} = \lambda^n e^{-\lambda \sum_{i=1}^{n} x_i}$$

The log-likelihood is

$$l = n \log(\lambda) - \lambda \sum_{i=1}^{n} x_i$$

and

$$\frac{dl}{d\lambda} = \frac{n}{\lambda} - \sum_{i=1}^{n} x_i$$

Solving $\dfrac{dl}{d\lambda} = 0$ gives $\hat{\lambda} = \dfrac{n}{\sum_{i=1}^{n} x_i} = \dfrac{1}{\overline{x}}$

3.5 CHOOSING BETWEEN ESTIMATORS

There often may be alternative estimators for the same unknown parameter; there are a number of criteria that may help to choose between them.

Bias: An unbiased estimator is one for which $E(\hat{\theta}) = \theta$. All other things being equal (which they rarely are) unbiased estimators are preferred.

Consistency: An estimator is consistent if it *converges in probability* to the true value of the parameter, where the term in italics means that the probability that $\hat{\theta}$ differs from θ by any given small quantity tends to zero as n tends to infinity. Estimators must be consistent.

Efficiency: The preferable estimator is the one that makes the most efficient use of the data in the sense that the distribution of the error $\hat{\theta} - \theta$ is as closely as possible concentrated near zero.

Maximum likelihood estimators are consistent and efficient but not always unbiased.

EXAMPLE 3.5

Is the maximum likelihood estimator of the variance of a normal distribution unbiased?

SOLUTION

The MLE estimator of σ^2 is the sample variance $s^2 = 1/n \sum_{i=1}^{n} (x_i - \bar{x})^2$, so we need to find the expected value of

$$E(s^2) = \frac{1}{n} \sum_{i=1}^{n} E\,(x_i - \bar{x})^2 = \frac{1}{n} E\left[\sum_{i=1}^{n}(x_i^2 - 2x_i\bar{x} + \bar{x}^2)\right]$$

$$= \frac{1}{n} E\left(\sum_{i=1}^{n} x_i^2 - n\bar{x}^2\right) = \frac{1}{n}\left[n(\sigma^2 + \mu^2) - n\left(\frac{\sigma^2}{n} + \mu^2\right)\right] = \sigma^2\left(1 - \frac{1}{n}\right)$$

So the MLE estimator is *not* unbiased in this case. An unbiased estimator for σ^2 is

$$s^2 = \frac{1}{n-1} \sum_{i=1}^{n} (x_i - \bar{x})^2$$

3.6 SAMPLING DISTRIBUTIONS: STUDENT'S *t*, CHI-SQUARE AND FISHER'S *F*

The *z*-statistic: The sampling distribution of the sample mean of samples from a population in which the random variable has a $N(\mu, \sigma^2)$ is $N(\mu, \sigma^2/n)$. Consequently a statistic, z, defined as

$$z = \frac{\bar{x} - \mu}{\sigma/\sqrt{n}}$$

will have a $N(0,1)$ sampling distribution. Some other sampling distributions of central importance in statistics are now defined below.

Student's *t*-distribution: If a random variable is distributed $N(\mu, \sigma^2)$ then the statistic t given by $t = \dfrac{(\bar{x} - \mu)}{(s/\sqrt{n})}$, calculated from a random sample of n values with mean $\bar{x}$ and standard deviation s, has what is known as *Student's t-distribution*; the distribution is given explicitly by

$$f(t) = \frac{\Gamma\left\{\frac{1}{2}(\upsilon+1)\right\}}{(\upsilon\pi)^{1/2}\Gamma\left(\frac{1}{2}\upsilon\right)}\left(1 + \frac{t^2}{\upsilon}\right)^{-\frac{1}{2}(\upsilon+1)} \quad \text{where } -\infty < t < \infty, \ \upsilon = n-1$$

The shape of the density function varies with υ and as υ gets larger its shape approaches that of a standard normal density (as might be guessed from comparing the definitions of the z and t statistics). Some examples of Student's *t*-densities are shown in Figure 3.3. The parameter υ is often called the *degrees of freedom* (df) of the density function. (NB: Degrees of freedom is an elusive concept that occurs throughout statistics. Essentially the term means the number of independent units of information in a sample relevant to the estimation of a parameter or calculation of a statistic.)

Chi-square distribution: If a random variable is distributed $N(\mu, \sigma^2)$ then the statistic $y = (n-1)s^2/\sigma^2$ calculated from a random sample of size n and where $s^2 = \dfrac{1}{n-1}\sum_{i=1}^{n}(x_i - \bar{x})^2$. (NB: this is the unbiased estimator of σ^2) has what is known as a *chi-square distribution* with $n-1$ degrees of freedom; the distribution is given by

$$f(y) = \frac{1}{2^{\upsilon/2}\Gamma(\upsilon/2)}y^{\upsilon-2}e^{-y/2}, y > 0, \upsilon = n-1$$

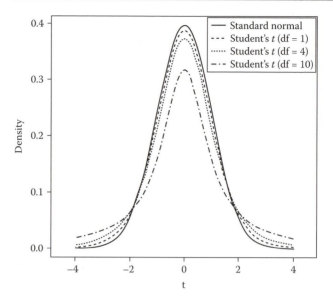

FIGURE 3.3 Student's *t*-distributions for different values of υ.

Some chi-square distributions with different degrees of freedom are shown in Figure 3.4. The chi-square distribution also arises as the distribution of the sum of squares of a number (υ) of independent standard normal variables. (NB: A random variable with a chi-square distribution is often denoted by the term χ^2.)

Fisher's *F*-distribution: If a random variable X has a $N(\mu_1, \sigma_1^2)$ and another random variable Y has a $N(\mu_2, \sigma_2^2)$, then the statistic $F = \dfrac{s_1^2}{\sigma_1^2} \Big/ \dfrac{s_2^2}{\sigma_2^2}$ where s_1^2 and s_2^2 are the unbiased estimators of the two population variances based on sample sizes of n_1 and n_2 for X and Y, respectively, has what is known as *Fisher's F-distribution* with $n_1 - 1$ and $n_2 - 1$ degrees of freedom; the distribution is given by

$$f(F) = \frac{\Gamma[(\upsilon_1 + \upsilon_2)/2]}{\Gamma(\upsilon_1/2)\Gamma(\upsilon_2/2)} \left(\frac{\upsilon_1}{\upsilon_2}\right)^{\upsilon_1/2} \frac{F^{\upsilon_1/2-1}}{\left[1 + (\upsilon_1 F/\upsilon_2)\right]^{(\upsilon_1/2+\upsilon_2/2)}},$$

$$F > 0, \upsilon_1 = n_1 - 1, \upsilon_2 = n_2 - 1$$

Some *F*-distributions with different degrees of freedom are shown in Figure 3.5.

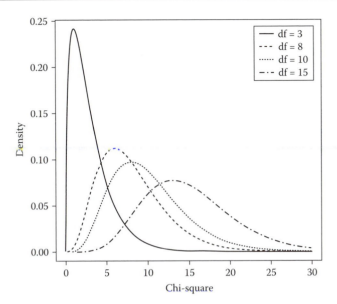

FIGURE 3.4   Chi-square distributions for different degrees of freedom.

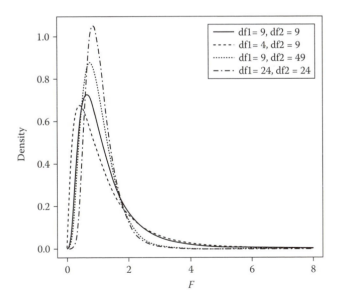

FIGURE 3.5   Fisher's *F*-distributions for various degrees of freedom.

(NB: The ratio of two χ^2 distributed variables, each divided by its degrees of freedom, has an F-distribution.)

3.7 INTERVAL ESTIMATION, CONFIDENCE INTERVALS

A point estimate of a parameter is of little use without some measure of its precision. Interval estimation, in particular, *confidence intervals* address this issue.
Confidence interval: A plausible range of values for an unknown parameter θ that is consistent with the sample data. Specifically the interval (L, U), where L and U are both random variables, is a $100(1-\alpha)\%$ confidence interval for θ if $\Pr(L \le \theta \le U) = 1 - \alpha$. The quantity $1 - \alpha$ is called the *confidence level* and is equal to the probability that the random interval (L, U) contains the fixed parameter θ. The confidence limits L and U are constructed from the observed data in such a way that in infinite replications of the study, the proportion of such intervals that contain the parameter θ is $1-\alpha$. (NB: If say 1.25 and 3.80 are the realized values of L and U for a given set of data, a statement such as $\Pr(1.25 < \theta < 3.80) = 0.95$ is statistically incorrect.)

EXAMPLE 3.6

How may a $100(1 - \alpha)\%$ confidence interval (CI) be constructed for the mean of a $N(\mu, \sigma^2)$ distribution where both parameters are unknown based on a random sample of n observations, $x_1, x_2, \ldots x_n$ from the distribution?

SOLUTION

First calculate the sample mean and sample variance (the unbiased version):

$$\bar{x} = \frac{1}{n} \sum_{i=1}^{n} x_i, s^2 = \frac{1}{n-1} \sum_{i=1}^{n} (x_i - \bar{x})^2$$

The sampling distribution of the statistic $t = \dfrac{(\bar{x} - \mu)}{(s/\sqrt{n})}$ is Student's t with $n-1$ degrees of freedom. So if $t_{n-1, 1-\alpha/2}$ is the $100(1 - \alpha/2)$ percentile point of Student's t-distribution with $n - 1$ degrees of freedom, and t_{obs} is the value of the t-statistic for the observed sample values, then

$$1-\alpha = \Pr(-t_{n-1,1-\alpha/2} < t_{obs} < t_{n-1,1-\alpha/2})$$

$$= \Pr\left(-t_{n-1,1-\alpha/2} < \frac{\bar{x}-\mu}{s/\sqrt{n}} < t_{n-1,1-\alpha/2}\right)$$

$$= \Pr(\bar{x} - t_{n-1,1-\alpha/2}s/\sqrt{n} < \mu < \bar{x} + t_{n-1,1-\alpha/2}s/\sqrt{n})$$

so that $L = \bar{x} - t_{n-1,1-\alpha/2}s/\sqrt{n}$ and $U = \bar{x} + t_{n-1,1-\alpha/2}s/\sqrt{n}$

Numerical example: A sample of 20 participants was drawn at random from a population and each measured on a psychological test assumed to have a normal distribution. The sample mean was 283.09 and the unbiased sample variance was 2644.02. For $\alpha = 0.05$, $t_{19,0.975} = 2.093$ (the required percentiles of the Student's t-distribution can be found either from tables or from your computer) which leads to the 95% CI for the mean being [259.02,307.16].

Other sampling distributions can be used to construct CIs for other parameters in a similar fashion.

EXAMPLE 3.7

How may a $100(1-\alpha)\%$ CI be constructed for the variance of a $N(\mu, \sigma^2)$ distribution based on a random sample of n values from the distribution?

SOLUTION

We first need the unbiased sample variance, s^2. The sampling distribtion of the statistic $y = (n-1)\dfrac{s^2}{\sigma^2}$ is chi-square with $n-1$ degrees of freedom. So if $c_{\alpha/2}$ and $c_{1-\alpha/2}$ are 'left-hand' and 'right-hand' critical points of the chi-square distribution with $n-1$ degrees of freedom then

$$1-\alpha = \Pr(c_{\alpha/2} \le y_{obs} \le c_{1-\alpha/2}) = \Pr\left(c_{\alpha/2} \le \frac{(n-1)s^2}{\sigma^2} \le c_{1-\alpha/2}\right)$$

$$= \Pr\left(\frac{(n-1)s^2}{c_{\alpha/2}} \le \sigma^2 \le \frac{(n-1)s^2}{c_{1-\alpha/2}}\right)$$

so $L = \dfrac{(n-1)s^2}{c_{\alpha/2}}$ and $U = \dfrac{(n-1)s^2}{c_{1-\alpha/2}}$

Numerical example: A random sample of 10 observations is drawn from a $N(\mu, \sigma^2)$ distribution and the unbiased sample variance is found to be 25. For $\alpha = 0.10$, $c_{\alpha/2} = 3.325$, $c_{1-\alpha/2} = 16.919$ and so a 90% CI for σ^2 is

$$\left[\frac{9 \times 25}{16.919}, \frac{9 \times 25}{3.325} \right] = [13.30, \ 67.67]$$

The CI is wide because the sample size is small.

In general, 95% and 99% CIs are constructed; as the confidence level increases so will the width of the CI.

(NB: It is the upper and lower limits of a confidence interval that are the random variables; the confidence level gives the probability that the random interval contains the value of the unknown parameter, *not* the probability that the unknown parameter lies in the interval. The parameter is *not* a random variable and so attaching probability statements to it is incorrect. Figure 3.6 makes this point graphically giving as it does the confidence intervals for the mean based on the *t* statistics given by 50 samples of size 20 from a N(5,1) distribution.)

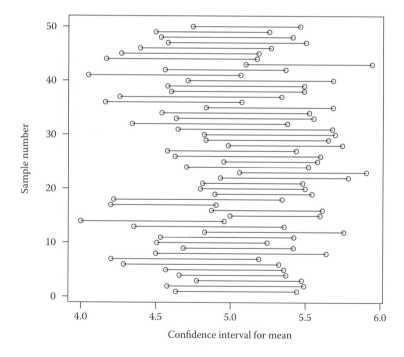

FIGURE 3.6 Student's *t* 95% CIs for the mean of a normal distribution each based on a sample of size 20 from a N(5,1) population.

3.8 SUMMARY

Point estimate: A single number calculated from a sample of values that provides an estimate of a population parameter.

Sampling distribution of the mean: The arithmetic mean calculated from a samples of n values from a $N(\mu, \sigma^2)$ has a $N(\mu, \sigma^2/\sqrt{n})$ distribution.

Central limit theorem: If a random variable X has population mean μ and population variance σ^2, the mean of a sample of n values from the population is approximately normally distributed with mean μ and variance σ^2/n for sufficiently large n.

Method of moments estimation: Estimation of unknown parameters by matching corresponding sample and population moments.

Maximum likelihood estimation: Estimation of unknown parameters by maximizing the likelihood of the data with respect to the parameter.

Properties of estimators:

- *Bias*: An unbiased estimator is one for which $E(\hat{\theta}) = \theta$. All other things being equal unbiased estimators are preferred.
- *Consistency*: An estimator is consistent if it *converges in probability* to the true value of the parameter, where the term in italics means that the probability that $\hat{\theta}$ differs from θ by any given small quantity tends to zero as n tends to infinity. Estimators must be consistent.
- *Efficiency*: The preferable estimator is the one that makes the most efficient use of the data in the sense that the distribution of the error, $\hat{\theta} - \theta$, is as closely as possible concentrated near zero.

Student's *t*-distribution: The probability distribution of the statistic $t = \dfrac{(\bar{x} - \mu)}{(s/\sqrt{n})}$ calculated from a sample of size n from a $N(\mu, \sigma^2)$.

Chi-square distribution: The probability distribution of the statistic $y = (n-1)s^2/\sigma^2$ calculated from a sample of size n from a $N(\mu, \sigma^2)$.

Fisher's *F*-distribution: The probability distribution of the statistic $$F = \frac{s_1^2/\sigma_1^2}{s_2^2/\sigma_2^2}$$

calculated from samples from two independent normal distributions, $N(\mu_1, \sigma_1^2)$ and $N(\mu_2, \sigma_2^2)$.

Confidence interval: A confidence interval for a unknown parameter θ represents a plausible range of values for the parameter that is consistent with the sample data.

Confidence interval for mean of a normal distribution:

$$\left[\bar{x} - t_{n-1,1-\alpha/2} s/\sqrt{n}, \bar{x} + t_{n-1,1-\alpha/2} s/\sqrt{n} \right]$$

Confidence interval for the variance of a normal distribution:

$$\left[\frac{(n-1)s^2}{c_{\alpha/2}}, \frac{(n-1)s^2}{c_{1-\alpha/2}} \right]$$

SUGGESTED READING

Bickel, PJ and Doksum, KA (2001) *Mathematical Statistics*, 2nd edition, Holden-Day, San Francisco.

Larsen, RJ and Marx, ML (2011) *Introduction to Mathematical Statistics and Its Applications*, Prentice Hall, Upper Saddle River, NJ.

Terrell, GR (2010) *Mathematical Statistics: A Unified Introduction*, Springer, New York.

Inference

4

4.1 INFERENCE AND HYPOTHESES

Statistical inference is the process of making conclusions on the basis of data that are governed by probabilistic laws. In essence inference is about testing the validity of claims, propositions or *hypotheses* made about a population of interest on the basis of a sample of data from that population.

Some examples of hypotheses:

- The average weight of men born in Essex and aged 67 in 2011 is 90 kg.
- Politicians in the Democratic and Republican parties have the same average IQ.
- Hair colour and eye colour of 15-year-old girls living in Paris are independent.
- The 10p coin I have in my pocket is 'fair', i.e., when tossed the probability of a head equals the probability of a tail.
- At age 10 boys in the UK are taller than girls age 10.

Hypotheses: In applying inferential methods two types of hypothesis are involved: the *null hypothesis*, H_0 and the *alternative hypothesis*, H_1 (sometimes called the *research hypothesis*). For example, when assessing whether a sample of weights of men born in Essex and aged 67 in 2011 was consistent with the average weight in the population being 90 kg, the two hypotheses might be

H_0: population average = 90 kg

H_1: population average $\neq$ 90 kg

More generally we might be interested in investigating whether some population parameter of interest, θ, takes some specified value, θ_0.

H_0: $\theta = \theta_0$

H_1: $\theta \neq \theta_0$

In some cases (the last bullet point above, for example) the alternative hypothesis may be *directional*:

H_0: $\theta = \theta_0$

H_1: $\theta < \theta_0$

or

H_0: $\theta = \theta_0$

H_1: $\theta > \theta_0$

Simple Hypothesis: A hypothesis that specifies the population distribution completely is a simple hypothesis; for example, H_0: The random variable $X \sim N(10,5)$.

Composite hypothesis: A hypothesis that does not specify the population distribution completely; for example, H_1: The random variable $X \sim N(\mu,5)$, $\mu > 10$.

4.2 SIGNIFICANCE TESTS, TYPE I AND TYPE II ERRORS, POWER AND THE z-TEST

Significance tests: Statistical procedures for evaluating the evidence from a sample of observations in respect to a null and an alternative hypothesis.

The logic behind a significance test is as follows:

- State the null and alternative hypotheses.
- Collect sample data that are informative about the hypotheses.
- Calculate the appropriate *test statistic*, the probability distribution of which is known under the null hypothesis; this will be the sampling distribution of the test statistic.

- From the sampling distribution of the test statistic under the null hypothesis calculate what is known as a *p-value*, which is defined as the probability of the observed data or data showing a more extreme departure from the null hypothesis *conditional* on the null hypothesis being true.
- Compare the *p*-value with a pre-specified *significance level*, α. If the *p*-value is less than α we say the sample data has produced evidence in favour of the alternative hypothesis and if the *p*-value is greater than α we conclude that there is no evidence against the null hypothesis.
- Conventionally a significance level of $\alpha = 0.05$ is used but this is only convention and other values might be more appropriate in some circumstances.
- Equivalent to comparing the *p*-value to α we can compare the observed value of the test statistic to the quantiles of the sampling distribution of the statistic which determine the *rejection region* of the test, i.e., the region(s) of the sampling distribution defining values of the test statistic that lead to evidence against the null hypothesis in favour of the alternative for the chosen significance level, α. The quantiles defining the rejection region are known as the *critical values*.
- All values of the test statistic in the rejection region will have associated *p*-values less than α.

Whenever a decision is reached about a hypothesis with a statistical test there is always a chance that our decision is wrong. We can quantify the uncertainty in terms of the probabilities of the two possible types of error.

RESULT FROM SIGNIFICANCE TEST	TRUE STATE OF THE WORLD	
	H_0 TRUE	H_1 TRUE
Evidence against H_0 (reject H_0 in favour of H_1)	Type I error Pr(Type I error) $= \alpha$	Correct decision Pr(that this occurs) $= 1 - \beta$ (known as *power* of the test)
No evidence against H_0 (accept H_0)	Correct decision	Type II error
	Pr(that this occurs) $= 1 - \alpha$	Pr(Type II error) $= \beta$

(NB: Although used in the above table and elsewhere in this section for convenience, it is better to avoid terms such as 'reject the null hypothesis' or 'accept the alternative hypothesis'; more appropriate terms are, for example, 'the test produces evidence against H_0 in favour of H_1' and 'the test gives no evidence against H_0'.)

(NB: If the significance level α is lowered from say 0.05 to say 0.01 to decrease the probability of a Type I error, then the probability of a Type II error will, of course, increase.)

EXAMPLE 4.1

In a population of interest it is known that IQ has a $N(100,20)$ distribution. The IQ of an individual is found to be 70. Is this score sufficiently below the mean of the population to doubt that the individual is from this population if we use a significance level of 0.05?

SOLUTION

The null hypothesis here is that the individual with an IQ of 70 *does* come from a $N(100,20)$ distribution. Under this null hypothesis we can calculate the probability that a score as low as 70 would be obtained for a member of the population. To find the required probability we simply convert the score to a standard normal value, $z = (70 - 100)/20 = -1.5$. The probability of a value from a standard normal distribution being less than or equal to 1.5 can be found from a table or from computer software and is 0.0668; as this is larger than 0.05 we conclude that there is no evidence that this individual is not from the putative population.

z-test: A very simple significance test for assessing hypotheses about the means of normal populations when the variance is known to be σ^2. The null hypothesis specifies a value for the mean say, H_0: $\mu = \mu_0$ (we shall consider the alternative hypothesis later). For a sample of size n, the test statistic is given by $z = (\bar{x} - \mu_0)\sqrt{n} / \sigma$ where $\bar{x}$ is the sample mean. The significance level at which we wish to apply the test is α. If H_0 is true, z has a standard normal distribution but how we use this fact to determine the p-value of the test depends upon the alternative hypothesis.

H_1: $\mu \neq \mu_0$

This is a non-directional hypothesis; if the test demonstrates that the null hypothesis is not acceptable then the findings may indicate either a larger *or* smaller value for the population mean. Observed values of the test statistic z that fall in *either* of the tails of the standard normal distribution as shown in Figure 4.1 lead to claiming evidence against the null hypothesis; the two tails are the rejection region of the z-test from this two-sided alternative hypothesis.

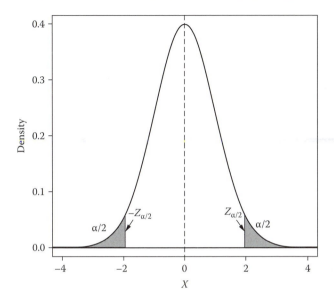

FIGURE 4.1   Rejection region for two-tailed z-test with significance level α; sampling distribution is the standard normal.

$H_1: \mu < \mu_0$ or $H_1: \mu > \mu_0$

These are both directional hypotheses and their rejection regions are the left-hand tail of the standard normal distribution ($\mu < \mu_0$) or the right-hand side of this distribution ($\mu > \mu_0$). These two rejection regions are shown in Figure 4.2.

EXAMPLE 4.2

A random sample of 16 observations is drawn from a normal distribution with unknown mean but known variance of 25. The arithmetic mean of the sample is 53. Use the z-test with significance level of 0.05 to assess the null hypothesis that the population mean is 50 against the alternative that it is not 50.

SOLUTION

In this example

$H_0: \mu = 50$

$H_1: \mu \neq 50$

The z-statistic takes the value $(53–50) \times 4/5 = 2.4$. For the 0.05 significance level the values defining the rejection regions for this alternative hypothesis are 1.96 and –1.96. The observed value of z is larger than 1.96 and therefore falls in the rejection region and we are led to conclude that there is evidence that the mean of the population is *not* 50. The associated p-value here is the probability that a standard normal variable is greater than 2.4 plus the probability that it is less than –2.4, giving the value 0.0164.

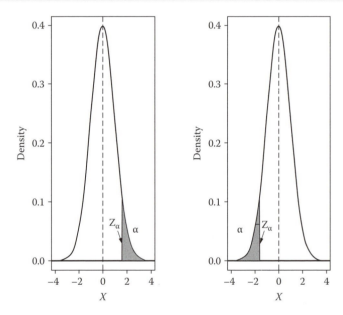

FIGURE 4.2 Rejection regions for one-tailed z-test with significance level α; sampling distribution is the standard normal.

4.3 POWER AND SAMPLE SIZE

Power: The probability of rejecting the null hypothesis when the alternative is true. Power is an important concept in choosing between alternative tests in some circumstances, with the test having the higher power for a given sample size being the best choice. Power is also important in planning a study and helping to answer the question, 'How large a sample do I need?'

Determining sample size to achieve a given power: We need first to specify the following quantities:

- Size of the Type I error, i.e., the significance level
- Likely variance of the response variable
- Power required
- A size of treatment effect that the researcher feels is important, i.e., a treatment difference that the investigators would not like to miss being able to declare to be statistically significant

Given such information the calculation of the corresponding sample size is often relatively straightforward, although the details will depend on the type of response variable and the type of test involved (see below for an example). In general terms the sample size will increase as the variability of the response variable increases and will decrease as the chosen clinically relevant treatment effect increases. In addition, the sample size will need to be larger to achieve a greater power and/or a more stringent significance level.

EXAMPLE 4.3

Derive a formula for estimating sample size when applying a z-test to test for a difference between the means of two normal populations each having variance σ^2, where the significance level is α and power $1 - \beta$ where β is the size of the Type II error. The sample sizes taken from each population are to be the same, and the alternative hypothesis is that one mean is greater than the other.

SOLUTION

$H_0: \mu_1 = \mu_2$

$H_1: \mu_1 > \mu_2$

Test statistic

$$z = \frac{\bar{x}_1 - \bar{x}_2}{\sigma\sqrt{\dfrac{1}{n} + \dfrac{1}{n}}}$$

Rejection region, values of test statistic greater than $z_{1-\alpha}$.
    The Type II error, β, is defined to be
$\beta = \Pr(\text{accepting } H_0 | H_1 \text{ is true}) = \Pr(z < z_{1-\alpha} | H_1 \text{ is true})$. If H_1 is true then

the test statistic z has a normal distribution with mean $\mu = \dfrac{\mu_1 - \mu_2}{\sigma\sqrt{\dfrac{2}{n}}}$ and variance equal to one. Therefore,

$$\beta = \Pr(z < z_{1-\alpha} | z \sim N(\mu,1)) = \int_{-\infty}^{z_{1-\alpha}-\mu} \frac{1}{\sqrt{2\pi}} e^{-x^2/2} dx$$

Also

$$\beta = \int_{-\infty}^{z_\beta} \frac{1}{\sqrt{2\pi}} e^{-x^2/2} dx$$

Consequently,

$z_\beta = z_{1-\alpha} - \mu$ leading to

$$\frac{(\mu_1 - \mu_2)}{\sigma\sqrt{\dfrac{2}{n}}} = z_{1-\alpha} - z_\beta \text{ and } n = \frac{2(z_{1-\alpha} - z_\beta)^2 \sigma^2}{(\mu_1 - \mu_2)^2}$$

Numerical example: Set $\sigma^2 = 16$, $\alpha = 0.05$, $\beta = 0.10$ so that power is 0.90 and $\mu_1 - \mu_2 = 1$, then $z_{0.95} = 1.64$ and $z_{0.10} = -1.28$ and the value of n is found to be 272.

4.4 STUDENT'S t-TESTS

The z-test is an unrealistic test for assessing hypotheses about the mean of a normal distribution because it assumes that the population variance is known rather than estimated from the sample data. Student's t-test overcomes this problem and can be used to test H_0: $\mu = \mu_0$ against a directional or unidirectional alternative. We assume we have a sample of n observations from a normal distribution with unknown mean and unknown variance; the test statistic is

$$t = \frac{\bar{x} - \mu_0}{s/\sqrt{n}}$$

(NB: This statistic involves s, the sample standard deviation, rather than the population value σ as in the z-test.)

When H_0 is true this statistic has as its sampling distribution a Student's t-distribution with $n - 1$ degrees of freedom. The rejection region of the test now involves the tail(s) of this distribution rather than the standard normal but otherwise the test is completely analogous to the z-test.

EXAMPLE 4.4

A random sample of 16 observations is drawn from a normal distribution with unknown mean and variance. The sample mean is 53 and the sample variance (the unbiased version) is 25. Use Student's t-test with a significance level of 0.05 to assess the hypothesis that the mean of the distribution is 50 against the alternative that it is not 50.

SOLUTION

In this example

$$H_0: \mu = \mu_0$$

$$H_1: \mu \neq \mu_0$$

The t-statistic takes the value $(53 - 50) \times 4/5 = 2.4$. For the 0.05 significance level the values defining the rejection region for this alternative hypothesis are found from a Student's t-distribution with 15 degrees of freedom to be 2.13 and –2.13. (NB: A larger value of the t-statistic than that of the z-statistic is needed to produce evidence against the null hypothesis, reflecting the extra uncertainty when estimating the population variances.) The observed value of t is in the rejection region and so the data produce evidence against the null hypothesis. The p-value here is the probability that a variable with a Student's t-distribution with 15 df is greater than 2.4 or less than –2.4, giving the value 0.03.

Two independent samples t-test: A significance test used to assess hypotheses about the equality of the means of two normal distributions. The test is based on random samples drawn from the two normal populations with unknown means and variances, although the two variances are assumed to be equal. The test statistic is

$$t = \frac{\bar{x}_1 - \bar{x}_2}{s\sqrt{\dfrac{1}{n_1} + \dfrac{1}{n_2}}}$$

where n_1 and n_2 are the sample sizes taken from each population, $\bar{x}_1$ and $\bar{x}_2$ are the respective sample means and s^2 is an estimate of the assumed common population variance given by

$$s^2 = \frac{(n_1 - 1)s_1^2 + (n_2 - 1)s_2^2}{n_1 + n_2 - 2}$$

where s_1^2 and s_2^2 are the respective unbiased sample variances.

The null hypothesis is

$$H_0: \mu_1 = \mu_2$$

and the most common alternative hypothesis is

$H_1: \mu_1 \neq \mu_2$ although others such as $H_1: \mu_1 < \mu_2$ might be of interest in particular circumstances.

Under H_0 the t-statistic has a Student's t-distribution with $n_1 + n_2 - 2$ degrees of freedom. The rejection regions are found from this distribution.

EXAMPLE 4.5

The data below give the percentage body fat for a sample of men and a sample of women. Use Student's independent sample t-test to assess the hypothesis that the population mean percentage body fat is the same in men and women against the alternative that the two means are not the same. What are the assumptions of the test?

Men percentage body fat: 9.5, 7.8, 17.8, 27.4
Women percentage body fat: 27.9, 31.4, 25.9, 25.2, 25.2

SOLUTION

In this example we have

$$H_0: \mu_1 = \mu_2$$

$$H_1: \mu_1 \neq \mu_2$$

$n_1 = 4, n_2 = 5, \bar{x}_1 = 15.6, \bar{x}_2 = 27.1, s_1^2 = 80.71, s_2^2 = 6.96, s^2 = 38.56,$
$t = -2.76.$

For significance level 0.05 the rejection region is values of the test statistic less than -2.36 or greater than 2.36. The observed value of the t-statistic lies in this region and we conclude that the data produce evidence against the null hypothesis. The p-value of the test statistic is 0.028. The assumptions of the test are that the observations are independent, that the population distribution of percentage body fat is normal in both men and women and that the two normal distributions have the same variance. (NB: The sizes of the two sample variances suggest that the equality of population variances assumption is suspect here.)

Paired samples t-test: Paired data can arise when a measurement is made on individuals on two occasions or when the individuals in two groups are individually matched. As the data of interest are now likely to be correlated rather than independent, the independent samples t-test is no longer appropriate for testing for a difference in means. The appropriate test now is the one sample t-test applied to the difference between the matched observations. The data now consist of n matched observations $(x_1, y_1), (x_2, y_2), \ldots, (x_n, y_n)$; the null hypothesis is $H_0: \delta = 0$ where δ is the population mean of the differences between matched observations (the alternative hypothesis is generally $H_1: \delta \neq 0$ although directional alternatives can be considered). The test statistic is

$$t = \frac{\bar{d}}{s_d / \sqrt{n}}$$

where s_d is the standard deviation of the sample differences $d_i = x_i - y_i$ and $\bar{d} = \dfrac{\sum\limits_{i=1}^{n} d_i}{n}$. If the null hypothesis is true then the statistic t has a Student's t-distribution with $n - 1$ degrees of freedom.

EXAMPLE 4.6

The measurement of supine systolic blood pressure was made on 10 patients with moderate essential hypertension immediately before and two hours after taking a drug, captopril. The data are

Patient:	1	2	3	4	5	6	7	8	9	10
Before	210	169	187	160	167	176	185	206	173	146
After	201	165	166	157	147	145	168	180	147	136

Is there any evidence that there has been a change in blood pressure?

SOLUTION

The appropriate test is the paired samples t-test; $\bar{d} = 16.7$, $s_d^2 = 95.57$, $t = 5.40$, p-value $= 0.00043$. There is very strong evidence against the null hypothesis and in favour of a change in blood pressure.

Assumptions: The single sample t-test is based on the assumption that the sample observations are from a population with a normal distribution. The paired t-test is based on the same assumption about the *differences* between the matched individuals. The independent sample t-test assumes that both samples are taken from populations with normal distributions and further assumes that the two normal distributions have the same variance.

Testing for normality using probability plots: The idea behind probability plotting is a simple one. We have a sample of observations, $x_1, x_2, \ldots, x_n$ from some population and we wish to know whether the sample values are compatible with the population distribution being normal. Rearrange the sample into ascending order as $x_{(1)} \leq x_{(2)} \leq \ldots \leq x_{(n)}$ where $x_{(i)}$ denotes the ith ordered observation. The ordered observations are plotted against the corresponding quantiles of the normal distribution y_i given by solving the equation

$$\Phi(y_i) = \frac{i}{n+1}$$

for $i = 1, 2, \ldots n$ where Φ represents the cumulative distribution function of the standard normal distribution, so that, for example, $\Phi(y_i) = \displaystyle\int_{-\infty}^{y_i} \frac{1}{\sqrt{2\pi}} e^{-\frac{1}{2}x^2} dx$. If the sample data are from a normal distribution then the points should lie approximately on a straight line.

EXAMPLE 4.7

Measurements of the chest measurement (in inches) were made on 20 individuals with the following results.

34.0, 37.0, 38.5, 36.5, 38.0, 43.0, 44.0, 39.5, 40.0, 41.0, 39.0, 32.0, 34.5, 33.0, 36.0, 37.0, 34.5, 32.5, 37.5, 35.0

Can we assume that the underlying population of chest measurements has a normal distribution?

SOLUTION

We construct a probability plot from the ordered observations and the corresponding normal quantiles.

Ordered observations: 32.0, 32.5, 33.0, 34.0, 34.5, 34.5, 35.0, 36.0, 36.5, 37.0, 37.0, 37.5, 38.0, 38.5, 39.0, 39.5, 40.0, 41.0, 43.0, 44.0
Normal quantiles: −1.67, −1.31, −1.07, −0.88, −0.71, −0.57, −0.43, −0.30, −0.18, −0.060, 0.060, 0.18, 0.30, 0.43, 0.57, 0.71, 0.88, 1.07, 1.31, 1.67

The probability plot is shown in Figure 4.3. There is no evidence of a departure from linearity so we can assume that the sample is from a population with a normal distribution.

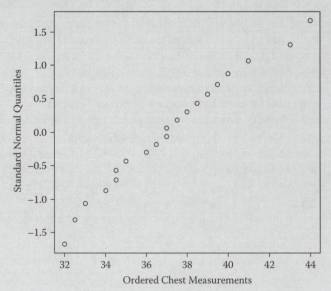

FIGURE 4.3 Normal probability plot for chest measurements.

Testing equality of variances for the independent samples *t*-test: The appropriate test is the *F*-test, which tests the null hypothesis, H_0: $\sigma_1^2 = \sigma_2^2$, against usually the alternative hypothesis, H_1: $\sigma_1^2 \geq \sigma_2^2$. The test statistic is the ratio of the larger of two sample variances, s_1^2 and s_2^2, to the smaller (we assume here that s_1^2 is the larger):

$$F = \frac{s_1^2}{s_2^2}$$

If H_0 is true then the sampling distribution of the test statistic is a Fisher's *F*-distribution with $n_1 - 1$ and $n_2 - 1$ degrees of freedom where n_1 and n_2 are the two sample sizes. The rejection region of the test is the right-hand tail of this *F*-distribution.

> **EXAMPLE 4.8**
> In Example 4.4 illustrating the independent samples *t*-test the two sample variances were 80.71 and 6.96. Is there any evidence that the two population variances differ?
>
> **SOLUTION**
> $F = 80.71/6.96 = 11.60$, $n_1 = 4$, $n_2 = 5$, *p*-value = 0.01, rejection region is values of the test statistic greater than 5.19. There is strong evidence that the variance of percentage fat in men is greater than the variance in women.

Aspin–Welch test: An alternative to the independent samples *t*-test when there is evidence that the two population variances are not equal. The test statistic is

$$t = \frac{\overline{x}_1 - \overline{x}_2}{\left(\dfrac{s_1^2}{n_1} + \dfrac{s_2^2}{n_2} \right)^{1/2}}$$

Under the equality of means hypothesis the statistic *t* has a Student's *t*-distribution with υ degrees of freedom where υ is the closest integer to

$$\frac{\left(\dfrac{s_1^2}{n_1} + \dfrac{s_2^2}{n_2} \right)^2}{\dfrac{s_1^4}{n_1^2(n_1 - 1)} + \dfrac{s_2^4}{n_2^2(n_2 - 1)}}$$

For the percentage fat data in Example 4.4 there is strong evidence that the variances of percentage fat differ in men and women, so we will apply the Aspin–Welch test to the data:

$t = -2.48$, $\upsilon = 3.42$, p-value $= 0.08$, rejection region $t < -2.97$ or $t > 2.97$

Using a two-sided Aspin–Welch test at a significance level of 0.05 we would conclude that there is no evidence that the means of percentage fat in men and women are different, a different conclusion from the independent samples t-test assuming equal population variances.

4.5 THE CHI-SQUARE GOODNESS-OF-FIT TEST

Chi-square goodness-of-fit test: A significance test for assessing the hypothesis that a sample of observations arises from a specified probability distribution. The test statistic is

$$X^2 = \sum \frac{(O_i - E_i)^2}{E_i}$$

where O_i and E_i are, respectively, the observed frequency and the expected frequency under the distributional hypothesis for variable value i and where the summation is taken over all values of the variable that have been observed. If the distributional hypothesis is correct, the test statistic has, *approximately*, a chi-square sampling distribution with $k - 1 - m$ degrees of freedom where k is the number of terms in the summation and m is the number of parameters of the hypothesised distribution that have to be estimated. The rejection region is the right-hand tail of this distribution. (NB: It is often suggested that the expected frequencies need to be greater than 5 to make the approximation that the chi-square distribution is, under the null hypothesis, the sampling distribution of X^2, acceptable, but this is often rather pessimistic and lower values may be acceptable.)

EXAMPLE 4.9

A die was tossed 20,000 times and the number of times each of the six faces appeared was as follows:

Face	1	2	3	4	5	6
Frequency	3407	3631	3176	2916	3448	3422

Was the die fair?

SOLUTION

The question asked can be translated into the simple distributional hypothesis that the probability of each face appearing when the die is rolled is 1/6. If this hypothesis is true the frequency to be expected for each face is the same and is 20,000/6 = 3333.333. The chi-square test statistic is therefore given by

$$X^2 = \frac{(3407-3333.333)^2}{3333.333} + \frac{(3631-3333.333)^2}{3333.333}$$

$$+ \frac{(3176-3333.333)^2}{3333.333} + \frac{(2916-3333.333)^2}{3333.333}$$

$$+ \frac{(3448-3333.333)^2}{3333.333} + \frac{(3422-3333.333)^2}{3333.333} = 94.19$$

If the die is fair, then X^2 will have a chi-square distribution with 5 degrees of freedom; the rejection region when testing at the 0.05 significance level is values of the test statistic greater than 11.07. Clearly there is very strong evidence that the die is not fair.

EXAMPLE 4.10

Data were collected on the emissions of α particles. There were 2612 equal time intervals during which emissions occurred and the results were as follows:

N of emissions	0	1	2	3	4	5	6	7	8	9	10	11	12	>12
Frequency	57	203	383	525	532	408	273	139	49	27	10	4	2	0

Does the number of emissions follow a Poisson distribution?

SOLUTION

The parameter of a Poisson distribution is estimated by the sample mean, here the value 3.877. So the hypothesis is that the number of α particles follows the Poisson distribution

$$\Pr(x \text{ emissions}) = \frac{3.877^x e^{-3.877}}{x!}$$

We have, therefore,

N OF EMISSIONS	POISSON PROBABILITY	EXPECTED FREQUENCY (2612 X PROBABILITY)
0	0.0207	54.1020
1	0.0803	209.7534
2	0.1557	406.6072
3	0.2012	525.4720
4	0.1950	509.3138
5	0.1512	394.9219
6	0.0977	255.1854
7	0.0541	141.3362
8	0.0262	68.4951
9	0.0113	29.5062
10	0.0044	11.4395
11	0.0015	4.0319
12	0.0005	1.3026
$>12\left(1 - \sum_{i=0}^{12} \Pr_i\right)$	0.0002	0.5326

Here $X^2 = 11.32$, df = 12 (one parameter has been estimated) for a significance level of 0.05 the rejection region is values of the test statistic greater than 21.03. There is no evidence against accepting that the number of emissions of α particles follows a Poisson distribution. (NB: As the expected values for counts of 11, 12 and > 12 are all less than 5 they should perhaps be amalgamated before using the chi-square goodness-of-fit test.)

4.6 NONPARAMETRIC TESTS

Nonparametric tests: Also known as *distribution free tests*, these are statistical tests that are based on a function of the sample observations, the probability distribution of which does not depend on a complete specification of the probability distribution of the population from which the sample is drawn. Consequently the tests are valid under relatively general assumptions about the underlying population. Such tests most often involve only the *ranks* of the observations rather than the observations themselves.

Wilcoxon–Mann–Whitney test: The nonparametric analogue to the two independent samples *t*-test based on ranking the observations, both samples combined, and then comparing the average ranks in the two samples. The test statistic is the sum of the ranks in the smaller sample (if the two sample sizes are equal the sum of the ranks in either group can be used). The hypothesis tested is that the two populations have different locations. For small sample sizes the *p*-value associated with the test statistic can be found from tables or will be given by the software package used. For larger sample sizes (above 25) the test statistic has, under the null hypothesis, an approximate normal distribution with mean equal to $n_1(n_1 + n_2 + 1)/2$ and variance $n_1 n_2 (n_1 + n_2 + 1)/12$.

EXAMPLE 4.11

Post-dexamethasone 1600-hour cortisol levels (in micrograms/dl) were recorded for a sample of psychiatric patients diagnosed with major depression and for another sample diagnosed with schizophrenia. Is there any evidence that the distribution of cortisol levels in the two diagnostic groups differs in location?

Major depression: 1.0,3.0,3.5,4.0,10.0,12.5,14.0,15.0,17.5,18.0,20.0, 21.0,24.5,25.0
Schizophrenia: 0.5,0.5,1.0,1.0,1.0,1.0,1.5,1.5,1.5,2.0,2.5,2.5,5.5,11.2

SOLUTION

Applying the Wilcoxon–Mann–Whitney test we find the value of the test statistic is $W = 179$, *p*-value = 0.0002056. There is very strong evidence of a difference in location of the cortisol levels of patients with major depression and those with schizophrenia.

Wilcoxon signed-rank test: The nonparametric analogue of the paired samples t-test based on the ranks of the absolute differences of the matched pairs of observations, with each rank being given the sign of the difference (zero differences are ignored). The test statistic is the sum of the positive ranks. The null hypothesis is that the probability distributions of the paired samples are the same. Exact p-values are available from tables or will be given by software. When the sample size is larger than 50 a normal approximation is available for the test statistic from knowing that its mean is $n(n+1)/4$ and its variance is $n(n+1)(2n+1)/24$.

EXAMPLE 4.12

The corneal thickness of the normal eye and the glaucomatous eye was measured for 8 patients with glaucoma. Is there any evidence that the probability distribution of corneal thickness differs in the two eyes?

PATIENT	1	2	3	4	5	6	7	8
Glaucomatous eye	488	478	480	426	440	410	458	460
Normal eye	484	478	492	444	436	398	464	476

SOLUTION

Applying the Wilcoxon signed-ranks test we find the test statistic to be $V = 7.5$, p-value $= 0.3088$. There is no evidence that the probability distribution of corneal thickness is different in the normal and glaucomatous eyes of patients with glaucoma.

4.7 TESTING THE POPULATION CORRELATION COEFFICIENT

Testing that the population correlation is zero: Under the assumption that the population correlation between two random variables, ρ, is zero, i.e., the two variables are not linearly related, then the test statistic, t, given by

$$t = r\sqrt{\frac{n-2}{1-r^2}}$$

where r is the sample correlation coefficient of the two variables and n is the sample size has a Student's t-distribution with $n-2$ degrees of freedom.

EXAMPLE 4.13

Assess whether or not the heights of husbands and their wives are independent using the sample heights of nine couples given below:

Husband height (mm)	1809	1841	1659	1779	1616	1695	1730	1753	1740
Wife height (mm)	1590	1560	1620	1540	1420	1669	1610	1635	1580

SOLUTION

We need to test the hypothesis that the population correlation coefficient between the heights of husbands and their wives is zero against the alternative that it is not zero, i.e.,

$$H_0: \rho = 0$$
$$H_1: \rho \neq 0$$

(directional alternatives might also be considered).

The sample correlation coefficient for the heights of the nine couples is 0.262, the t-statistic takes the value 0.72, the p-value is 0.50 and the rejection region is values of r greater than 2.36 or less than –2.36. There is no evidence against the null hypothesis that the population correlation of the heights of husbands and their wives is zero.

Testing that the correlation coefficient takes a specified non-zero value: To test the null hypothesis H_0: $\rho = \rho_0$ where $\rho_0 \neq 0$ the test statistic is based on Fisher's z-transformation of the sample correlation coefficient, i.e.,

$$z = \frac{1}{2} \log\left(\frac{1+r}{1-r}\right)$$

When the null hypothesis is true, z has a normal distribution with mean $\mu = \frac{1}{2} \log\left(\frac{1+\rho_0}{1+\rho_0}\right)$ and variance $\sigma^2 = 1/(n-3)$. We can now use the test statistic Z given by

$$Z = \frac{z - \mu}{\sigma}$$

which under H_0 has a standard normal distribution to test the null hypothesis. This test is based on the assumption that in the population the two variables have a *bivariate normal distribution* (see Chapter 10).

EXAMPLE 4.14

Final examination scores (out of 75) and corresponding exam completion times (seconds) were collected for 10 individuals taking the same examination. Test the hypothesis that the population correlation between examination scores and completion times takes the value –0.6 using a significance level of 0.05.

Subject	1	2	3	4	5	6	7	8	9	10
Exam score	49	49	70	55	52	55	61	65	57	71
Completion time	2860	2063	2013	2000	1420	1934	1519	2735	2329	1590

SOLUTION

The null hypothesis is H_0: $\rho = -0.6$ with H_1: $\rho \neq -0.6$ (directional alternatives could be used).

$r = -0.19$, $z = -0.19$, $\mu = -0.69$, $\sigma^2 = 0.14$, $Z = 1.33$, p-value = 0.18, rejection region $Z > 1.96$, $Z < -1.96$

The data provide no evidence against the null hypothesis.

4.8 TESTS ON CATEGORICAL VARIABLES

Testing the equality of two proportions: Binary variable observed in two populations:

H_0: $\pi_1 = \pi_2$
H_1: $\pi_1 \neq \pi_2$

where π_1 and π_2 are the population proportions of, say, the zero category of the binary variable. From samples of size N_1 and N_2 from the two populations; calculate the two sample proportions, $\hat{p}_1 = \dfrac{n_1}{N_1}$ and $\hat{p}_2 = \dfrac{n_2}{N_2}$, where n_1 and n_2 are the number of individuals in the zero category in each sample; these are estimates of the corresponding population proportions. Under H_0 the estimate of the proportion of the zero category is $\hat{p} = \dfrac{n_1 + n_2}{N_1 + N_2}$. The test statistic is

$$z = \frac{\hat{p}_1 - \hat{p}_2}{\sqrt{\hat{p}(1 - \hat{p})\left(\dfrac{1}{n_1} + \dfrac{1}{n_2}\right)}}$$

Under H_0 z has a standard normal distribution.

EXAMPLE 4.15

Fifty individuals are treated with a drug and 50 with a placebo; 15 in the drug group and 4 in the placebo group suffer nausea. Test the hypothesis that the proportions of nausea sufferers in the treatment and placebo populations are the same against the alternative that they differ. Use a significance level of 0.05.

SOLUTION

Here we have $p_1 = 15/50 = 0.30$ and $p_2 = 4/50 = 0.80$. The value of the test statistic is $z = 2.80$. The rejection region is values of the test statistic greater than 1.96 or less than -1.96. There is strong evidence against the null hypothesis. A 95% confidence interval for the difference in proportions can be calculated as

$$(p_1 - p_2) \pm 1.96 \sqrt{\frac{p_1(1-p_1)}{n_1} + \frac{p_2(1-p_2)}{n_2}}$$

leading to [0.07, 0.37].

Testing the equality of two proportions with matched samples: Binary variable observed in two populations that are matched.

$H_0: \pi_1 = \pi_2$

$H_1: \pi_1 \neq \pi_2$

where π_1 and π_2 are the population proportions of, say, the zero category of the binary variable. Here the sample data consist of n paired observations and these can be divided into four groups according to whether the zero category of the variable is present or not in each member of the pair to give the following frequencies:

yes/yes pair: a
yes/no pair: b
no/yes pair: c
no/no pair: d
with $n = a + b + c + d$

The test statistic is

$$z = \frac{b - c}{\sqrt{b + c}}$$

Under H_0 this has a standard normal distribution. (NB: $z^2 = \dfrac{(b-c)^2}{b+c}$ and could be tested as a chi-square with 1 df. This is known as *McNemar's test*.)

EXAMPLE 4.16

A group of 32 marijuana users were compared to 32 matched controls with respect to their sleeping difficulties. The results were

SLEEP DIFFICULTIES

MARIJUANA GROUP	CONTROL GROUP	FREQUENCY
Yes	Yes	4
Yes	No	3
No	Yes	9
No	No	16
Total		32

Is there any evidence of a difference in the proportion of individuals with sleeping difficulties in the two groups?

SOLUTION

The test statistic is $z = -1.73$; the rejection region for the 0.05 significance level and a two-sided alternative hypothesis is $z > 1.96$ and $z < -1.96$. There is insufficient evidence to reject the null hypothesis.

Chi-square test of the independence of two categorical variables: We have two categorical variables, one with r categories and one with c categories and N individuals have been cross-classified with respect to the two variables to give the following contingency table:

		VARIABLE 1				
		1	2	...	c	TOTAL
	1	n_{11}	n_{12}	...	n_{1c}	$n_{1.}$
	2	n_{21}	n_{22}	...	n_{2c}	$n_{2.}$
Variable 2	⋮	⋮	⋮	⋮	⋮	⋮
	r	n_{r1}	n_{r2}	...	n_{rc}	$n_{r.}$
	Total	$n_{.1}$	$n_{.2}$	...	$n_{.c}$	n

The null hypothesis of interest is that the two variables are independent, which can be formulated as follows:

$$H_0\colon p_{ij} = p_{i.}\,p_{.j} \text{ with } H_1\colon p_{ij} \neq p_{i.}\,p_{.j}$$

where p_{ij} is the probability of an observation falling in the ijth cell of the cross-classification table and $p_{i.}$ and $p_{.j}$ are, respectively, the probabilities of falling in the ith category of variable 2 irrespective of variable 1 and of falling into the jth category of variable 1 irrespective of variable 2. Under the null hypothesis the estimated expected value in the ijth cell is given by

$$E_{ij} = \frac{n_{i.}\,n_{.j}}{N}$$

and the test statistic is

$$X^2 = \sum_{i=1}^{r}\sum_{j=1}^{c} \frac{\left(n_{ij} - E_{ij}\right)^2}{E_{ij}}$$

If H_0 is true then the test statistic has a chi-square distribution with $(r-1)(c-1)$ degrees of freedom. The rejection region is the right-hand tail of the appropriate chi-square distribution.

EXAMPLE 4.17

The table below gives the counts of responses of people from three groups (normal, mild psychiatric illness, severe psychiatric illness) to the question 'Have you recently found that taking your own life kept coming into your mind?'

CATEGORY	DEFINITELY NOT	I DON'T THINK SO	HAS CROSSED MY MIND	DEFINITELY HAS
Normal	90	5	3	1
Mild psychiatric illness	43	18	21	15
Severe psychiatric illness	34	8	21	36

Test whether the two variables are independent using a significance level of 0.05.

SOLUTION

The null hypothesis here is that thoughts about suicide and illness group are independent. The estimated expected values under this hypothesis are as follows:

CATEGORY	DEFINITELY NOT	I DON'T THINK SO	HAS CROSSED MY MIND	DEFINITELY HAS
Normal	56.04	10.40	15.01	17.45
Mild psychiatric illness	54.91	10.19	14.80	17.10
Severe psychiatric illness	56.04	10.40	15.10	17.45

The test statistic takes the value 91.25; the p-value is very, very small. The rejection region is values of the test statistic greater than 12.59. There is very strong evidence that the two variables are not independent.

2 × 2 contingency table: A cross classification of two binary variables. The general 2 × 2 table may be written in terms of the observed frequencies from a sample of N individuals:

		VARIABLE 1		
		0	1	TOTAL
Variable 2	0	a	b	$a + b$
	1	c	d	$c + d$
	Total	$a + c$	$b + d$	$a + b + c + d = N$

Here the test statistic for assessing the independence of the two variables can be written:

$$X^2 = \frac{N(ad - bc)^2}{(a+b)(c+d)(a+c)(b+d)}$$

and is tested as a chi-square with a single degree of freedom.

EXAMPLE 4.18

For the 2 × 2 contingency table given below calculate the value of the chi-square statistic. (These are the data used in Example 4.15.)

| | NAUSEA | | |
	PRESENT	ABSENT	TOTAL
Drug given	15	35	50
Placebo	4	46	50

SOLUTION

The chi-square statistic takes the value 7.86. (NB: The chi-square statistic is the square of the z-statistic in Example 4.15; the two tests are equivalent. For a 2×2 contingency table, testing for independence is equivalent to testing the equality of two population proportions.)

Fisher's test: A test for the independence of two categorical variables for sparse data based on the hypergeometric distribution (see Chapter 1). Most commonly applied to a 2×2 table. Assuming the two variables are independent, the probability, P, of obtaining any arrangement of the frequencies a, b, c and d when the marginal totals are as given is

$$P = \frac{(a+b)!(a+c)!(c+d)!(b+d)!}{a!b!c!d!N!}$$

Fisher's test calculates P and the probabilities of other arrangements of the frequencies deviate more from the null hypothesis of independence when the marginal totals are fixed at their values in the original table. The sum of these probabilities is the p-value of the test.

EXAMPLE 4.19

A study was carried out to investigate the causes of jeering or baiting behaviour by a crowd when a person is threatening to commit suicide by jumping from a high building. A hypothesis is that baiting is more likely to occur in hot weather. The data collected are given in the following 2×2 table (the data are from the Northern Hemisphere).

| | BEHAVIOUR | |
	BAITING	NON-BAITING
June–September	8	4
October–May	2	7

Is there any evidence of an association between baiting and warm weather?

SOLUTION

Using Fisher's exact test we find the p-value to be 0.08. There is no strong evidence of crowd behaviour being associated with time of year of threatened suicide.

4.9 THE BOOTSTRAP

Bootstrap methods are procedures for the empirical estimation of sampling distributions and can be used in the construction of confidence intervals or tests of hypotheses when the form of the population distribution for a random variable is unknown. The bootstrap approach involves sampling with replacement from the original sample values.

EXAMPLE 4.20

The following 10 values are a random sample from a standard normal distribution:

0.303, −1.738, 0.208, −0.148, 0.136, 0.360, −0.588, 1.827, −1.359, 0.201

The true sampling distribution of the mean of samples of size 10 from the standard normal is N(0,0.1). Use the bootstrap approach to estimate the sampling distribution and compare the estimated distribution with the true sampling distribution.

SOLUTION

We sample the 10 observations with replacement 2000 times and use the 2000 mean values to construct an approximation to the sampling distribution of the sample mean. A histogram of the approximate sampling distribution is given in Figure 4.4. Also shown in the figure is the true N(0,0.1) sampling distribution. The approximation is clearly very good.

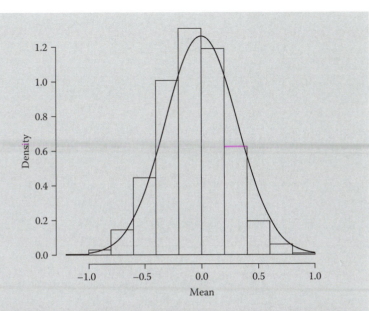

FIGURE 4.4 Approximate sampling distribution of the sample mean produced by bootstrapping 10 observations from a standard normal distribution and the true N(0,0.1) distribution.

EXAMPLE 4.21

The following 10 observations are a random sample from a standard normal distribution:

   −1.500, −0.126, 0.001, −0.794, −0.095, 0.192, 2.193, −0.612, 0.009, 0.587

Use the bootstrap approach to find an empirical sampling distribution of the sample median and from this extract a 95% confidence interval for the median.

SOLUTION

We sample the 10 observations with replacement 2000 times and use the 2000 values of the median as an approximation to the sampling distribution; this is shown in Figure 4.5. The 95% CI extracted from this is [−0.703,0.158].

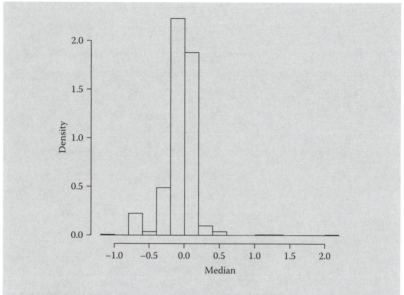

FIGURE 4.5 Approximate sampling distribution of the sample median produced by bootstrapping 10 observations from a standard normal distribution.

4.10 SIGNIFICANCE TESTS AND CONFIDENCE INTERVALS

Significance tests and confidence intervals are essentially two sides of the same coin. All values of a parameter lying within the limits of a $(1 - \alpha)$ confidence interval will be values which, when specified in the null hypothesis and tested at significance level α, will lead to the null hypothesis being 'accepted'. Conversely all values of the parameter outside the limits of the confidence interval will, when used as the null hypothesis value of the parameter, lead to 'rejection' of the null hypothesis in favour of the alternative.

The p-value from a significance test represents only limited information about the results from a study. The excessive use of p-values in hypothesis testing, simply as a means of rejecting or accepting a particular hypothesis at the expense of other ways of assessing results, has reached such a degree that levels of significance are often quoted alone in the main text and abstracts of papers with no mention of other more relevant and important quantities. The implications of hypothesis testing—that there can always be a simple 'yes' or 'no' answer as the fundamental result from a research study—are clearly false and, used in this way, hypothesis testing is of limited value.

Estimating the value of a parameter along with some interval that includes the population value of the parameter with a specified probability that the interval contains the 'true' value of the parameter, i.e., a confidence interval, gives a plausible range of values for the unknown parameter. The significance test relates to what the population parameter is not; the confidence interval, gives a plausible range for what the parameter is.

So, should the p-value be abandoned completely? Many statisticians would (grumpily) answer yes, but I think a more sensible response, at least for many applied researchers, would be a resounding 'maybe'. The p-value should rarely be used in a purely confirmatory way, but in an exploratory fashion, p-values can be useful in giving some informal guidance on the possible existence of an interesting effect even when the required assumptions of whatever test is being used are known to be invalid.

4.11 FREQUENTIST AND BAYESIAN INFERENCE

Frequentist inference: The basic idea that lies behind frequentist inference is that it is possible to consider an infinite sequence of independent repetitions of the same statistical experiment. Hypothesis testing, significance tests and confidence intervals are part of the frequentist school of inference, still the most widely used method of inference in practice.

Bayesian inference: A second school of inference based on Bayes' theorem that differs from frequentist inference is the use of the investigator's knowledge about the parameters of interest before seeing the data; such knowledge is summarized by what is known as the *prior distribution*. There is no equivalent to the prior distribution in frequentist inference.

The principal steps in Bayesian inference are

1. Formulate the prior distribution, $P(\theta)$, expressing what is known about the parameter(s), θ, prior to collecting any data.
2. Find the likelihood of the data, D, in terms of the unknown parameters, $P(D|\theta)$.

3. Apply Bayes' theorem to derive the posterior distribution, $p(\theta/D)$ which expresses what is known about the parameters after observing the data:

$$P(\theta|D) = \frac{P(\theta)P(D|\theta)}{\int P(\theta)P(D|\theta)}$$

Quantities calculated from this posterior distribution of the parameters form the basis of inference. Point estimates of parameters may be found by calculating the mean, median or mode of a parameter posterior distribution, with parameter precision being estimated by the standard deviation or some suitable inter-quantile range, for example, from quantiles at p and $1 - p$ for a $100(1 - 2p)\%$ *credible interval* for a parameter. In general, such quantities, $f(\theta)$, will be estimated by their posterior expectation, given by

$$E[f(\theta)|D] = \frac{\int f(\theta)P(\theta)P(D|\theta)d\theta}{\int P(\theta)P(D|\theta)d\theta}$$

(When there is more than a single parameter the integration in the above equation will be, of course, a multiple integration.)

EXAMPLE 4.22
Consider a sequence of Bernoulli trials where the single parameter, θ, is the probability of a 'success'. We will assume that we can characterise any prior knowledge we have as to the likely value of θ in a particular type of prior distribution, namely, a beta distribution with density function

$$P(\theta) = \frac{\Gamma(\alpha+\beta)\theta^{\alpha-1}(1-\theta)^{\beta-1}}{\Gamma(\alpha)\Gamma(\beta)}, \ 0 \le \theta \le 1$$

This density function has mean $\alpha/(\alpha + \beta)$ and variance $\alpha \beta/[(\alpha + \beta)^2 (\alpha + \beta + 1)]$. What is the posterior distribution?

SOLUTION
In a sequence of n Bernoulli trials in which we observe r successes the likelihood is proportional to $\theta^r(1-\theta)^{n-r}$. The posterior distribution is obtained by multiplying the prior distribution by the likelihood to give

$$P(\theta \mid r,n) \propto \theta^{\alpha+r+1}(1-\theta)^{\beta+n-r-1}$$

which is another beta distribution with parameters $\alpha = \alpha + r$ and $\beta^* = \beta + n-r$. The mean of the posterior distribution is $(\alpha + r)/(\alpha + \beta + n)$; as α and β approach zero, corresponding to a prior distribution with the greatest possible variance, so the posterior mean approaches r/n, the value that would be expected under maximum likelihood. As r and n increase relative to α and β so the variance of the distribution approaches the familiar, $\dfrac{(r/n)\,(1-r/n)}{n}$. (The beta distribution is what is known as the *conjugate prior* for the Bernoulli parameter, which simply means that the posterior distribution is also a beta distribution as we have shown.)

4.12 SUMMARY

Statistical inference: The process of making conclusions on the basis of data that are governed by probabilistic laws.

Hypotheses*:* In applying inferential methods two types of hypothesis are involved: the *null hypothesis*, H_0, and the *alternative hypothesis*, H_1.

Significance tests: Statistical procedures for evaluating the evidence from a sample of observations with respect to a null and an alternative hypothesis.

Test statistic: A statistic calculated from sample data, the probability distribution (sampling distribution) of which is known under the null hypothesis.

p-value: The probability of the observed data or data showing a more extreme departure from the null hypothesis *conditional* on the null hypothesis being true.

Rejection region of a test: The region(s) of the sampling distribution defining values of the test statistic that leads to evidence against the null hypothesis in favour of the alternative.

Power: The probability of rejecting the null hypothesis when the alternative is true.

z-test: A significance test for assessing hypotheses about the means of normal populations when the variance is known to be σ^2. For a sample of size n the test statistic is given by $z = (\bar{x} - \mu_0)\sqrt{n} / \sigma$

where $\bar{x}$ is the sample mean. The sampling distribution of z is the standard normal distribution.

One sample Student's t-test: A test of H_0: $\mu = \mu_0$ against a directional or unidirectional alternative. The test statistic is $t = \dfrac{\bar{x} - \mu_0}{(s/\sqrt{n})}$. The sampling distribution of t is a Student's-t distribution with $n - 1$ degrees of freedom.

Two independent samples t-test: A significance test used to assess hypotheses about the equality of the means of two normal distributions. The test statistic is

$$t = \frac{\bar{x}_1 - \bar{x}_2}{s\sqrt{\dfrac{1}{n_1} + \dfrac{1}{n_2}}}$$

where s^2 is an estimate of the assumed common population variance given by

$$s^2 = \frac{(n_1 - 1)s_1^2 + (n_2 - 1)s_2^2}{n_1 + n_2 - 2}$$

Paired samples t-test: The appropriate test is the one sample t-test applied to the difference between the matched observations. The test statistic is $t = \dfrac{\bar{d}}{(s_d/\sqrt{n})}$.

Assumptions: The single sample t-test is based on the assumption that the sample observations come from a population with a normal distribution. The paired t-test is based on the same assumption about the *differences* between the matched individuals. The independent sample t-test assumes that both samples are taken from populations with normal distributions and further assumes that the two normal distributions have the same variance.

Fisher's F-test: Tests the null hypothesis, H_0: $\sigma_1^2 = \sigma_2^2$, usually against the alternative hypothesis, H_1: $\sigma_1^2 \geq \sigma_2^2$. The test statistic is the ratio of the larger of two sample variances, s_1^2 and s_2^2, to the smaller (we assume here that s_1^2 is the larger $F = \dfrac{s_1^2}{s_2^2}$. The sampling distribution of F is Fisher's F with $n_1 - 1$ and $n_2 - 1$ degrees of freedom.

Aspin–Welch test: An alternative to the independent samples t-test when there is evidence that the two population variances are not equal. The test statistic is $t = \dfrac{\bar{x}_1 - \bar{x}_2}{\left(\dfrac{s_1^2}{n_1} + \dfrac{s_2^2}{n_2}\right)^{1/2}}$. Under the equality of means hypothesis the statistic t has a Student's t distribution with υ degrees of freedom where υ is the closest integer to

$$\upsilon = \frac{\left(\dfrac{s_1^2}{n_1} + \dfrac{s_2^2}{n_2}\right)^2}{\dfrac{s_1^4}{n_1^2(n_1 - 1)} + \dfrac{s_2^4}{n_2^2(n_2 - 1)}}$$

Chi-square goodness-of-fit test: A significance test for assessing the hypothesis that a sample of observations arises from a specified probability distribution. The test statistic is $X^2 = \sum \dfrac{(O_i - E_i)^2}{E_i}$. The sampling distribution is a chi-square distribution with $k - m - 1$ degrees of freedom.

Nonparametric tests: Statistical tests that are based on a function of the sample observations, the probability distribution of which does not depend on a complete specification of the probability distribution of the population from which the sample is drawn.

Wilcoxon–Mann–Whitney test: The nonparametric analogue to the two independent samples t-test based on ranking the observations, both samples combined, and then comparing the average ranks in the two samples. The test statistic is the sum of the ranks in the smaller sample (if the two sample sizes are equal, the sum of the ranks in either group can be used).

Wilcoxon signed-rank test: The nonparametric analogue of the paired samples t-test based on the ranks of the absolute differences of the matched pairs of observations, with each rank being given the sign of the difference (zero differences are ignored). The test statistic is the sum of the positive ranks.

Testing H_0: $\rho = 0$: The test statistic is $t = r\sqrt{\dfrac{n-2}{1-r^2}}$ where r is the sample correlation coefficient of the two variables and n is the sample size. The sampling distribution of t is a Student's t-distribution with $n - 2$ degrees of freedom.

Testing H_0: $\rho = \rho_0$ where $\rho \neq \rho_0$. The test statistic is based on Fisher's z-transformation of the sample correlation coefficient, i.e., $z = \frac{1}{2}\log\frac{1+r}{1-r}$. When the null hypothesis is true z has a normal distribution with mean $\mu = \frac{1}{2}\log\frac{1+\rho_0}{1+\rho_0}$ and variance $\sigma^2 = 1/(n-3)$.

Testing equality of proportions:

Test statistic $z = \dfrac{\hat{p}_1 - \hat{p}_2}{\sqrt{\hat{p}(1-\hat{p})(1\backslash n_1 + 1\backslash n_2)}}$

Testing independence of two categorical variables: Test statistic $X^2 = \sum_{i=1}^{r}\sum_{j=1}^{c}\dfrac{(n_{ij}-E_{ij})^2}{E_{ij}}$.

Fisher's exact test: $P = \dfrac{(a+b)!(a+c)!(c+d)!(b+d)!}{a!b!c!d!N!}$

Bootstrap methods: Procedures for the empirical estimation of sampling distributions and can be used in the construction of confidence intervals or tests of hypotheses when the form of the population distribution for a random variable is unknown. The bootstrap approach involves sampling with replacement from the original sample values.

Frequentist inference: The basic idea that lies behind frequentist inference is that it is possible to consider an infinite sequence of independent repetitions of the same statistical experiment. Hypothesis testing, significance tests and confidence intervals are part of the frequentist school of inference, which is still the most widely used method of inference in practice.

Bayesian inference: A second school of inference based on Bayes' theorem that differs from frequentist inference in the use of the investigator's knowledge about the parameters of interest before seeing the data. Such knowledge is summarised by what is known as the *prior distribution*. There is no equivalent to the prior distribution in frequentist inference.

SUGGESTED READING

Howell, DC (2002) *Statistics in Psychology*, Duxbury, Pacific Grove, CA.

Larsen, RJ and Marx, ML (2011) *Introduction to Mathematical Statistics and Its Applications*, Prentice Hall, Upper Saddle River, NJ.

van Belle, G, Fisher, LD, Heagerty, PJ and Lumley, T (2004) *Biostatistics,* 2nd edition, Wiley, New York.

Analysis of Variance Models

5

5.1 ONE-WAY ANALYSIS OF VARIANCE

Analysis of variance: The separation of variance ascribable to one group of causes from the variance ascribable to other groups. Such separation leads to suitable tests for the equality of the means of several populations.

One-way analysis of variance: The simplest example of the analysis of variance; used to test the hypothesis that the population means of a variable of interest in the k categories of a categorical variable (a *factor* variable in this context) are equal, based on independent random samples of observations from each category. The model assumed for the observations is

$$y_{ij} = \mu + \alpha_i + \varepsilon_{ij}$$

where y_{ij} is the value of the variable of interest for the jth observation in the ith level of the factor variable, $i = 1,\ldots,k; j = 1,\ldots,n_i$ and n_i is the number of sample values in the ith category, $i = 1,\ldots,k$; μ represents the overall mean of the observations, α_i quantifies how much the mean of the ith category deviates from the overall mean and ε_{ij} represents an error term which we assume is $N(0, \sigma^2)$. The model implies that $E(y_{ij}) = \mu + \alpha_i$ and $Var(y_{ij}) = \sigma^2$; in addition, $\sum_{i=1}^{ni} \alpha_i = 1$. The equality of means hypothesis can be translated into the following null hypothesis about α_i:

H$_0$: $\alpha_1 = \alpha_2 = \ldots = \alpha_k = 0$.
H$_1$: At least one α_i is not equal to zero.

Here the total variation in the observations can be separated into parts due to the variation between group means and the variation of the observations within groups (variances are referred to here as *mean squares*). The process can be summarized in an *analysis of variance table.*

SOURCE	SUM OF SQUARES (SS)	DEGREES OF FREEDOM (DF)	MEAN SQUARES (MS)	EXPECTED MEAN SQUARES
Between groups	$BSS = \sum_{i=1}^{k} n_i(\bar{y}_{i.} - \bar{y}_{..})^2$	$k - 1$	$BMS = BSS/(k-1)$	$\dfrac{\sum_{i=1}^{k} n_i\alpha_i^2}{k-1} + \sigma^2$
Within groups	$WSS = \sum_{i=1}^{k}\sum_{j=1}^{n_i}(y_{ij} - \bar{y}_{i.})^2$	$n - k$	$WMS = WSS/(n-k)$	σ^2
Total	$TSS = \sum_{i=1}^{k}\sum_{j=1}^{n_i}(y_{ij} - \bar{y}_{..})^2$	$n - 1$		

($n = \Sigma_{i=1}^{k} n_i$, TSS = WSS + BSS: The '.' nomenclature is used to indicate summation over a subscript.)

Under H_0 both the between groups and within groups mean squares are estimators of the variance σ^2. Under H_1 the between groups mean square estimates a larger variance term. Consequently, the test statistic for assessing the null hypothesis is the mean square ratio, MSR = BMS/WMS, which if H_0 is true has an F-distribution with $k - 1$ and $n - k$ degrees of freedom.

EXAMPLE 5.1

The silver content (%Ag) of a number of Byzantine coins discovered in Cyprus was determined; the coins came from four different periods in the 12th century. Is there any evidence of a difference in silver content minted in the different periods?

PERIOD 1	PERIOD 2	PERIOD 3	PERIOD 4
5.9	6.9	4.9	5.3
6.8	9.0	5.5	5.6
6.4	6.6	4.6	5.5
7.0	8.1	4.5	5.1
6.6	9.3		6.2
7.7	9.2		5.8
7.2	8.6		5.8
6.9			
6.2			

SOLUTION

It is always good practice to start any data analysis with an informative graphic; here a boxplot is used (Figure 5.1).

The plot suggests that there are differences between the average silver content of coins minted in different periods and perhaps also suggests that the variance of the observations in the second period is larger than the variance in the other three periods.

The analysis of variance table for these data is

SOURCE	SS	DF	MS	F = MSR
Between periods	37.75	3	12.58	26.27
Within periods	11.02	23	0.48	

If we use the 0.05 significance level, the rejection region is values of the MSR greater than 3.03; the *p*-value here is very, very, small. There is strong evidence against the equality of means hypothesis although this does *not* imply that all four period means differ.

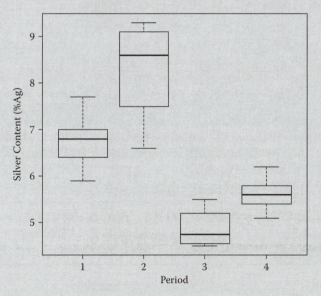

FIGURE 5.1 Boxplots of silver content of coins minted in different periods.

The assumptions made for the F-tests to be valid are that the population period variances are equal and that the error terms or residuals in the model have a normal distribution. We can assess the latter assumption informally with a probability plot of the sample residuals, $y_{ij} - y_i$ (Figure 5.2).

There is some departure from linearity in the probability plot but not enough to cause concern given the tiny p-value associated with the F-test.

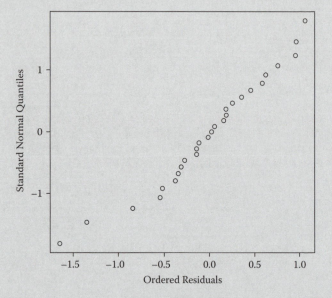

FIGURE 5.2   Probability plot for sample residuals for silver content of coins.

5.2 FACTORIAL ANALYSIS OF VARIANCE

Factorial designs: Studies in which observations on a variable of interest are collected for all combinations of a number of factor variables. The simplest factorial design is one involving two factors. In such studies the effects of each of the factors and *interactions* between the factors can be assessed.

Balanced two-way analysis of variance: For a two-factor design with factors A (with levels 1,2,...r) and B (with levels 1,2,...c) the model used is as follows:

$$y_{ijk} = \mu + \alpha_i + \beta_j + (\alpha\beta)_{ij} + \varepsilon_{ijk}$$

where y_{ijk} represents the kth observation in the jth level of factor B and the ith level of factor A, μ is the overall mean, α_i is the effect of the ith level of factor A (*main effect* of A), β_j is the effect of the jth level of factor B (main effect of B) and $(\alpha\beta)_{ij}$ is the *interaction effect* of the two factors in the ijth factor combination. The ε_{ijk} are the ubiquitous independent error terms assumed to be N(0, σ^2). The model is formulated in such a way that $\sum_{i=1}^{r}\alpha_i = 0$; $\sum_{j=1}^{c}\beta_j = 0$; $\sum_{i=1}^{r}(\alpha\beta)_{ij} = 0$; $\sum_{j=1}^{c}(\alpha\beta)_{ij} = 0$.

Finally, in a balanced design the number of observations in each combination of levels of the two factors is the same and this number will be represented by n.

Interaction effect: Measures the difference between the effect of A and B together and the sum of their main effects.

The null hypotheses of interest in the two-way design are

$$H_0^{(1)}: \alpha_1 = \alpha_2 = \ldots = \alpha_r = 0;$$

$$H_0^{(2)}: \beta_1 = \beta_2 = \ldots = \beta_c = 0;$$

$$H_0^{(3)}: (\alpha\beta)_{ij} = 0, i = 1,\ldots,r; j = 1,\ldots,c.$$

In each case the alternative hypothesis is that at least one of the effects is zero. The analysis of variance table here is as follows.

SOURCE	SUM OF SQUARES (SS)	DEGREES OF FREEDOM (DF)	MEAN SQUARES (MS)	MEAN SQUARE RATIOS	EXPECTED MEAN SQUARE
Between levels of A	$SSA = n\sum_{i=1}^{r}(\bar{y}_{i..} - \bar{y}_{...})^2$	$r-1$	$MSA = \dfrac{SSA}{r-1}$	$MSRA = \dfrac{MSA}{MSW}$	$\sigma^2 + \dfrac{n\sum_{i=1}^{r}\alpha_i^2}{r-1}$
Between levels of B	$SSB = n\sum_{j=1}^{c}(\bar{y}_{.j.} - \bar{y}_{...})^2$	$c-1$	$MSB = \dfrac{SSB}{c-1}$	$MSRB = \dfrac{MSB}{MSW}$	$\sigma^2 + \dfrac{\sum_{j=1}^{c}\beta_j^2}{c-1}$
Interaction A × B	$SSAB = n\sum_{i=1}^{r}\sum_{j=1}^{c}(\bar{y}_{ij.} - \bar{y}_{i..} - \bar{y}_{.j.} + \bar{y}_{...})^2$	$(r-1)(c-1)$	$MSAB = \dfrac{SSAB}{(r-1)(c-1)}$	$MSRAB = \dfrac{MSAB}{MSW}$	$\sigma^2 + \dfrac{n\sum_{i=1}^{r}\sum_{j=1}^{c}(\alpha\beta)_{ij}^2}{(r-1)(c-1)}$
Within cells	$SSW = \sum_{i=1}^{r}\sum_{j=1}^{c}\sum_{k=1}^{n}(y_{ijk} - \bar{y}_{ij.})^2$	$rc(n-1)$	$MSW = \dfrac{SSW}{rc(n-1)}$		σ^2
Total	$TSS = \sum_{i=1}^{r}\sum_{j=1}^{c}\sum_{k=1}^{n}(y_{ijk} - \bar{y}_{...})^2$	$rcn-1$			

Note: TSS = SSA + SSB + SSAB + SSW

Under $H_0^{(1)}$, MSA and MSW are both estimators of σ^2. Under $H_1^{(1)}$, MSA estimates a larger variance and, consequently, we can use an F-distribution with $r-1$ and $rc(n-1)$ degrees of freedom to test the hypothesis. For the other two MSRs there are corresponding F-tests to assess the relevant hypotheses.

EXAMPLE 5.2

Data were collected in an investigation into types of slimming regimes. The two-factor variables were the type of advice given to a slimmer with two levels, 'psychological' and 'non-psychological' and status of the slimmer, again with two levels, 'novice' and 'experienced'. Weight loss in pounds over three months was recorded for four women in each of the four possible categories giving the following data:

Status	ADVICE Non-Psychological	ADVICE Psychological
Novice	2.85, 1.98, 2.12, 0.00	4.44, 8.11, 9.40, 3.50
Experienced	2.42, 0.00, 2.74, 0.84	0.00, 1.64, 2.40, 2.15

Carry out a two-way analysis of variance of the data.

SOLUTION

The analysis of variance table for the data is

SOURCE	SS	DF	MS	MSR	p-VALUE
Between status	25.53	1	25.53	8.24	0.014
Between advice	21.83	1	21.83	7.04	0.021
Status x advice	20.95	1	20.95	6.76	0.023
Within cells	37.19	12	3.10		

Here both main effects and the interaction are highly significant; interpretation of the significant interaction is helped by the plot in Figure 5.3.

The clear message here is

1. Non-psychological advice, no difference in weight loss between experienced and novice slimmers.
2. Psychological advice novice slimmers win.

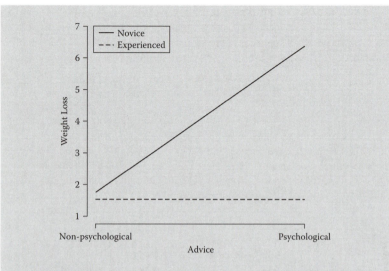

FIGURE 5.3 Graphic display of the interaction of status and advice; the four cell means are plotted.

Unbalanced two-way analysis of variance: When the number of observations in the cells of a two-way design are unequal it is said to be unbalanced; now there is no way of finding a 'sum of squares' corresponding to each main effect and their interaction because the effects are no longer independent. It is now no longer possible to partition the total variation in the response variable into non-overlapping or orthogonal sums of squares representing factor main effects and factor interactions. For example, there is a proportion of the variance of the response variable that can be attributed to (explained by) either factor A or factor B, and so, consequently, A and B together explain less of the variation of the response than the sum of which each explains alone. The result of this is that the sum of squares that can be attributed to a factor depends on which factors have already been allocated a sum of squares; in other words, the sums of squares of factors and their interactions depend on the *order* in which they are considered.

EXAMPLE 5.3

Rats were fed four diets generated by the crossing of two factors, protein source (levels, beef and cereal) and protein amount (levels, low and high).

(NB: The weight gained by each rat over a month was recorded in grams. A different number of rats received each diet.)

	PROTEIN AMOUNT	PROTEIN AMOUNT
Protein source	Low	High
Beef	90, 76, 90, 64, 86	73,102,118
Cereal	107, 95, 97, 80	98, 74, 56, 111, 95, 88, 82

Carry out a two-way analysis of variance of the data.

SOLUTION

The data are unbalanced so we need to consider the main effects in two different orders to give the following two-way analysis of variance tables:

SOURCE	SS	DF	MS	MSR	p-VALUE
Source of protein	18.3	1	18.2	0.07	0.79
Amount of protein/source of protein	19.7	1	19.7	0.08	0.79
Source x amount/ source and amount	671.1	1	671.1	2.63	0.13
Within cells	3827.6	15	255.2		

SOURCE	SS	DF	MS	MSR	p-VALUE
Amount of protein	29.1	1	29.1	0.11	0.74
Source of protein/ amount of protein	8.9	1	8.9	0.04	0.85
Source x amount/ source and amount	671.1	1	671.1	2.63	0.13
Within cells	3827.6	15	255.2		

The sums of squares now essentially represent the *extra* sum of squares when comparing the addition of a term to a current model; for example, source of protein *given that* amount of protein is already in the model.

5.3 MULTIPLE COMPARISONS, A PRIORI AND POST HOC COMPARISONS

A priori comparisons: Also known as *planned comparisons* in which a specific hypothesis is made about a set of population means and then tested, rather than using more general hypotheses. (NB: Planned comparisons have to be just that and not the results of hindsight after inspection of the sample means.) A planned comparison is usually formulated in terms of a particular *contrast* between the means being zero where for k population means a contrast is defined as

$$C = c_1\mu_1 + c_2\mu_2 + \ldots + c_k\mu_k \text{ where } \sum_{i=1}^{k} c_i = 0.$$

The null hypothesis is that $C = 0$. The estimate of the contrast is obtained from the sample means as

$$L = c_1\bar{x}_1 + c_2\bar{x}_2 + \ldots + c_k\bar{x}_k$$

The sum of squares of the contrast is

$$SS_{contrast} = \frac{L^2}{\sum_{i=1}^{k} \frac{c_i^2}{n_i}}$$

This has a single degree of freedom and the contrast sum of squares is tested against the within group mean square with the rejection region being the right-hand tail of the appropriate F-distribution.

EXAMPLE 5.4

Data were collected in an investigation into the effect of the stimulant caffeine on the performance of a simple task. Forty male students were trained in finger tapping. They were then divided randomly into four groups of 10 and groups received different doses of caffeine (0, 100, 200 and 300 ml). Two hours after treatment each man was required to carry out finger tapping and the number of taps per minute was recorded.

0 ML	100 ML	200 ML	300 ML
242	248	246	248
245	246	248	250
244	245	250	251
248	247	252	251
247	248	248	248
248	250	250	251
242	247	252	252
244	246	248	249
246	243	245	253
242	244	250	251

Test H_0: $\mu_0 - \frac{1}{3}\mu_{100} - \frac{1}{3}\mu_{200} - \frac{1}{3}\mu_{300} = 0$

SOLUTION

The sample means are $\bar{x}_0 = 244.8$, $\bar{x}_{100} = 246.4$, $\bar{x}_{200} = 248.3$, $\bar{x}_{300} = 250.4$. The estimate of the contrast of interest is $244.8 - 1/3 \times 246.4 - 1/3 \times 248.3 - 1/3 \times 250.4 = -3.57$. The associated sum of squares is

$$\frac{(-3.57)^2}{1/10[1+(1/3)^2+(1/3)^2+(1/3)^2]} = 95.41$$

The within group mean square for the data is 4.40 and so the MSR for the contrast is $95.41/4.40 = 21.68$. Testing this against an F-distribution with 1 and 36 degrees of freedom the p-value is found to be 0.000043. There is very strong evidence that the specified contrast differs from zero; more finger tapping takes place when subjects are given caffeine.

Post hoc comparisons of means: A significant F-test in an analysis of variance indicates that not all the population means are equal. But it does not tell us which means are different from which other means. Post hoc tests (*unplanned comparisons*) allow a more detailed investigation of the mean differences.

A question: Why not test for a difference between each pair of means using the usual independent samples t-test? This is why:

$$H_0: \mu_1 = \mu_2 = \ldots = \mu_k$$

The number of t-tests needed to compare each pair of means is $N = k(k-1)/2$. Suppose each t-test is performed with significance level α so that for each of the tests Pr(rejecting the equality of the two means given that they are equal) = α.

Consequently, Pr(accepting the equality of the two means when they are equal) = $1 - \alpha$. Therefore, Pr(accepting the equality for all N of the t-tests performed when H_0 is true) = $(1 - \alpha)^N$. And so finally, Pr(rejecting the equality of at least one pair of means when H_0 is true) = $1 - (1 - \alpha)^N = P$ (say).

For particular values of k and for $\alpha = 0.05$ the corresponding values of P are given below.

K	N	P
2	1	0.05
3	3	0.14
4	6	0.26
10	45	0.90

The probability of falsely rejecting the null hypothesis quickly increases above the nominal significance level.

There are a variety of procedures (*multiple comparison tests*) that can be used to investigate mean differences whilst trying to keep the overall significance level close to the nominal level.

Fisher's least significant difference procedure: The application of independent sample t-tests to each pair of means but *only after* an F-test has rejected the equality of means hypothesis. (NB: Use not advised when a large number of means are being compared.)

Bonferonni procedure: The application of independent sample t-tests to each pair of means but using a significance level of α/m rather than α for each of the m tests performed. Alternatively (and preferably) a confidence interval can be constructed as

$$\text{Difference in sample means } \pm t_{n-k,1-\alpha/2m}\, s\sqrt{1/n_1 + 1/n_2}$$

where s^2 is the within group mean square from the analysis of variance, n is the total number of observations, k is the number of categories of the factor variable and n_1 and n_2 are the number of observations in the two levels being compared. (NB: Can be very conservative.)

Scheffe's procedure: Again independent sample t-tests are the basis of the procedure but here each observed test statistic is tested against $\left[(k-1)F_{k-1,n-k}(\alpha)\right]^{1/2}$. Again a confidence interval can be constructed as

Difference in sample means $\pm \left[(k-1)F_{k-1,n-k}(\alpha) \right]^{1/2} \times s\sqrt{1/n_1 + 1/n_2}$

(NB: Very conservative for a small number of comparisons.)

EXAMPLE 5.5

Returning to the silver content of coins data used in Example 5.1 we can calculate the confidence intervals for differences between each pair of means using each of the procedures described above from the four group means, 6.74, 8.24, 4.88, 5.61 and the value of s found from the analysis of variance table as $\sqrt{0.4789} = 0.692$.

Fisher's LSD procedure: The analysis of variance has produced a significant F value giving strong evidence against the equality of means hypothesis. We can therefore calculate the usual t-tests for differences between pairs of means; here we give the corresponding 95% confidence intervals.

Period 1-Period 2: $(6.74-8.24) \pm 0.692 \times 2.07 \times \sqrt{1/9+1/7} = [-2.22, -0.78]$

Period 1-Period 3: $(6.74-4.88) \pm 0.692 \times 2.07 \times \sqrt{1/9+1/4} = [1.15, 2.59]$

Period 1-Period 4: $(6.74-5.61) \pm 0.692 \times 2.07 \times \sqrt{1/9+1/7} = [0.41, 1.85]$

Period 2-Period 3: $(8.24-4.88) \pm 0.692 \times 2.07 \times \sqrt{1/7+1/4} = [2.65, 4.09]$

Period 2-Period 4: $(8.24-5.61) \pm 0.692 \times 2.07 \times \sqrt{1/7+1/7} = [1.91, 3.35]$

Period 3-Period 4: $(4.88-5.61) \pm 0.692 \times 2.07 \times \sqrt{1/4+1/7} = [-1.46, -0.02]$

Bonferonni procedure: Here the 95% confidence intervals are

Period 1-Period 2: $(6.74-8.24) \pm 0.692 \times 2.89 \sqrt{1/9+1/7} = [-2.50, -0.49]$

Period 1-Period 3: $(6.74-4.88) \pm 0.692 \times 2.89 \sqrt{1/9+1/4} = [0.86, 2.88]$

Period 1-Period 4: $(6.74-5.61) \pm 0.692 \times 2.89 \sqrt{1/9+1/7} = [0.12, 2.14]$

Period 2-Period 3: $(8.24-4.88) \pm 0.692 \times 2.89 \sqrt{1/7+1/4} = [2.36, 4.37]$

Period 2-Period 4: $(8.24-5.61) \pm 0.692 \times 2.89 \sqrt{1/7+1/7} = [1.62, -3.64]$

Period 3-Period 4: $(4.88-5.61) \pm 0.692 \times 2.89 \sqrt{1/4+1/7} = [-1.75, 0.27]$

Scheffe procedure: Here the confidence intervals are

Period 1-Period 2: $(6.74 - 8.24) \pm 0.692 \times 3.01 \sqrt{1/9 + 1/7}$ = [-2.55, -0.45]

Period 1-Period 3: $(6.74 - 4.88) \pm 0.692 \times 3.01 \sqrt{1/9 + 1/4}$ = [0.82, 2.92]

Period 1-Period 4: $(6.74 - 5.61) \pm 0.692 \times 3.01 \sqrt{1/9 + 1/7}$ = [0.08, 2.18]

Period 2-Period 3: $(8.24 - 4.88) \pm 0.692 \times 3.01 \sqrt{1/7 + 1/4}$ = [2.32, 4.42]

Period 2-Period 4: $(8.24 - 5.61) \pm 0.692 \times 3.01 \sqrt{1/7 + 1/7}$ = [1.58, 3.68]

Period 3-Period 4: $(4.88 - 5.61) \pm 0.692 \times 3.01 \sqrt{1/4 + 1/7}$ = [-1.79, 0.31]

5.4 NONPARAMETRIC ANALYSIS OF VARIANCE

When the normality assumption needed for the F-test used in the one-way analysis of variance is suspect, an alternative nonparametric method that can be used is the *Kruskal–Wallis test*. The null hypothesis is now that the k-populations have the same median and the alternative is that at least two populations have different medians. The observations are assumed to be independent and there should be no tied values. The combined sample observations from all k groups are ranked and the test statistic is

$$KW = \frac{12}{n(n+1)} \sum_{i=1}^{k} \frac{R_i^2}{n_i} - 3(n+1) \text{ where } n = \sum_{i=1}^{k} n_i$$

and R_i is the rank-sum of the ith group.

For large sample sizes the sampling distribution of KW when the null hypothesis is true is chi-square with $k - 1$ degrees of freedom; the rejection region is the right-hand tail of this distribution. The large sample approximation requires a minimum group size of 7 when $k = 3$ and 14 when $k = 6$ for significance level 0.05.

EXAMPLE 5.6

In an investigation of the possible beneficial effects of pre-therapy training of clients on the process and outcome of counselling and psychotherapy, four methods were tried and nine clients assigned to each. A measure of psychotherapeutic attraction was given to each client at the end of the training period.

CONTROL	THERAPEUTIC READING	VIDEOTAPED PRE-TRAINING	GROUP, DIRECT LEARNING
0	0	0	1
1	6	5	5
3	7	8	12
3	9	9	13
5	11	11	19
10	13	13	22
13	20	16	25
17	20	17	27
26	24	20	29

Is there any difference in the average psychotherapeutic score between the four training methods?

SOLUTION

Here we will apply the Kruskal–Wallis test. The test statistic has a value of 4.26 with 3 degrees of freedom. The p-value found from the appropriate chi-square distribution is 0.23. There is no evidence of differences in the medians of the four training method populations.

5.5 SUMMARY

One-way analysis of variance: Independent samples of observations from each of k populations, sample sizes, $n_1, n_2, ..., n_k$, total sample size, $n = \sum_{i=1}^{k} n_i$. Model for data

$$y_{ij} = \mu + \alpha_i + \varepsilon_{ij}; \ \varepsilon_{ij} \sim N(0, \sigma^2); \ \sum_{i=1}^{k} \alpha_i = 0$$

$H_0: \alpha_1 = \alpha_2 = ... = \alpha_k = 0,$

$H_1:$ at least one non-zero α_i.

MSR = MS between groups/MS within groups: If H_0 is true MSR has a F-distribution with $k - 1$ and $n - k$ degrees of freedom. The critical region is values of MSR $> F_{k-1,n-k}(\alpha)$ where α is the chosen significance level.

Two-way analysis of variance: For a two-factor design with factors A (with levels r) and B (with levels c) the model is as follows:

$$y_{ijk} = \mu + \alpha_i + \beta_j + (\alpha\beta)_{ij} + \varepsilon_{ijk}; \varepsilon_{ijk} \sim N(0,\sigma^2);$$

$$\sum_{i=1}^{r}\alpha_i = 0; \sum_{j=1}^{c}\beta_j = 0; \sum_{i=1}^{r}(\alpha\beta)_{ij} = \sum_{j=1}^{c}(\alpha\beta)_{ij} = 0.$$

F-tests for main effects of A and B and for their interaction.

Planned comparisons: Tests of contrasts of population means,

$$c = c_1\mu_1 + c_2\mu_2 + \ldots + c_k\mu_k \text{ where } \sum_{i=1}^{k}c_i = 0.$$

H_0: $C = 0$.

$L = c_1\bar{x}_1 + c_2\bar{x}_2 + \ldots + c_k\bar{x}_k$, s^2 is the within cell mean square.

$$SS_{contrast} = \frac{L^2}{\displaystyle\sum_{i=1}^{k}\frac{c_i^2}{n_i}}$$

$F = SS_{contrast}/s^2$, test against and F-distribution with $1, n - k$ df.

Multiple comparisons: Tests between each pair of population means.

Fisher's least significant difference procedure: After the F-test has rejected to equality of population means, a confidence interval for each pair of means is

$$(\bar{x}_i - \bar{x}_j) \pm t_{n-k,1-\alpha/2}\, s\sqrt{1/n_i + 1/n_j}$$

Bonferonni procedure: Use a significance level of α/m rather than α for each of the m tests performed. Confidence interval is

$$\left(\bar{x}_i - \bar{x}_j\right) \pm t_{n-k,1-\alpha/2m}\ s\sqrt{1/n_i + 1/n_j}$$

Scheffe's procedure: Confidence interval given by

$$\left(\bar{x}_i - \bar{x}_j\right) \pm \left[(k-1)F_{k-1,n-k}(\alpha)\right]^{1/2} s\sqrt{1/n_1 + 1/n_2}$$

Kruskal–Wallis test. The null hypothesis is now that the k-populations have the same median and the alternative is that at least two populations have different medians. The test statistic is

$$KW = \frac{12}{n(n+1)}\sum_{i=1}^{k}\frac{R_i^2}{n_i} - 3(n+1) \text{ where } n = \sum_{i=1}^{k}n_i$$

and R_i is the rank-sum of the ith group.

For large sample sizes the sampling distribution of KW when the null hypothesis is true is chi-square with $k-1$ degrees of freedom; the rejection region is the right-hand tail of this distribution.

SUGGESTED READING

Howell, DC (2002) *Statistics in Psychology*, Duxbury, Pacific Grove, CA.
Roberts, M and Russo, R (1999) *A Student's Guide to Analysis of Variance*, Routledge, New York.

Linear Regression Models

6

6.1 SIMPLE LINEAR REGRESSION

Regression: A frequently applied statistical technique that serves as a basis for studying and characterizing a system of interest, by formulating a reasonable mathematical model of the relationship between a response variable y and a set of p explanatory variables $x_1, x_2, \ldots x_p$. The choice of an explicit form of the model may be based on previous knowledge of a system or on considerations such as 'smoothness' and continuity of y as a function of the explanatory variables (sometimes called the *independent variables,* although they are rarely independent; explanatory variables is the preferred term).

Simple linear regression: A linear regression model with a single explanatory variable. The data consist of n pairs of values $(y_1, x_1), (y_2, x_2), \ldots (y_n, x_n)$. The model for the observed values of the response variable is

$$y_i = \beta_0 + \beta_1 x_i + \varepsilon_i, \ i = 1 \ldots n$$

where β_0 and β_1 are, respectively, the *intercept* and *slope* parameters of the model and the ε_i are error terms assumed to have a $N(0, \sigma^2)$ distribution. The parameters β_0 and β_1 are estimated from the sample observations by *least squares*, i.e., the minimization of $S = \sum_{i=1}^{n} \varepsilon_i^2$

$$S = \sum_{i=1}^{n} (y_i - \beta_0 - \beta_1 x_i)^2$$

$$\frac{\partial S}{\partial \beta_0} = -2 \sum_{i=1}^{n} (y_i - \beta_0 - \beta_1 x_i)$$

$$\frac{\partial S}{\partial \beta_1} = -2 \sum_{i=1}^{n} (y_i - \beta_0 - \beta_1 x_i) x_i$$

Setting $\dfrac{\partial S}{\partial \beta_0} = 0$, $\dfrac{\partial S}{\partial \beta_1} = 0$ leads to the following estimators of the two model parameters:

$$\hat{\beta}_0 = \bar{y} - \hat{\beta}_1 \bar{x}, \ \hat{\beta}_1 = \frac{\displaystyle\sum_{i=1}^{n} (y_i - \bar{y})(x_i - \bar{x})}{\displaystyle\sum_{i=1}^{n} (x_i - \bar{x})^2}$$

The variance σ^2 is estimated by $s^2 = \dfrac{\displaystyle\sum_{i=1}^{n} (y_i - \bar{y})^2}{n-2}$. The estimated variance of the estimated slope parameter is $\mathrm{Var}(\hat{\beta}_1) = \dfrac{s^2}{\displaystyle\sum_{i=1}^{n} (x_i - \bar{x})^2}$. The estimated variance of a predicted value y_{pred} at a given value of x, say, $x_{0,}$ is

$$\mathrm{Var}(y_{\mathrm{pred}}) = s^2 \sqrt{1 + \frac{1}{n} + \frac{(x_0 - \bar{x})^2}{\displaystyle\sum_{i=1}^{n} (x_i - \bar{x})^2}}$$

EXAMPLE 6.1

Ice cream consumption (pints per capita) was measured over ten 4-week periods and the mean temperature (degrees F) was also recorded over the same times.

Consumption	0.386	0.374	0.393	0.425	0.406	0.344	0.327	0.288	0.269	0.256
Temperature	41	56	63	68	69	65	61	47	32	24

Find the linear regression equation linking consumption to temperature.

SOLUTION

$\hat{\beta}_0 = 0.192, \hat{\beta}_1 = 0.003, s^2 = 0.0016$

Confidence interval for β_1 : $\hat{\beta}_1 \pm t_{n-1,1-\alpha/2} \sqrt{\text{Var}(\hat{\beta}_1)}$: [0.001,0.005]

Each increase in temperature of one degree F is estimated to increase sales of ice cream between 0.001 and 0.005 pints per capita. A plot of the data and the fitted regression line are shown in Figure 6.1.

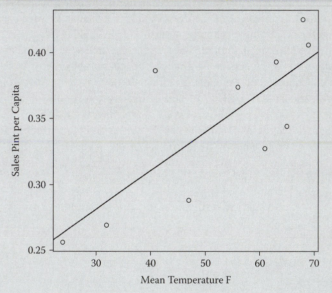

FIGURE 6.1   Scatterplot of ice cream sales against mean temperature showing fitted regression line

6.2 MULTIPLE LINEAR REGRESSION

Multiple linear regression: A generalization to more than a single explanatory variable of the simple linear regression model. The multiple linear regression model is given by

$$y_i = \beta_0 + \beta_1 x_{i1} + \cdots + \beta_p x_{ip} + \varepsilon_i$$

y_i represents the value of the response variable on the ith individual, and $x_{i1}, x_{i2}, ..., x_{ip}$ represent this individual's values on p explanatory variables, with $i = 1, 2, ... n$. As usual, n represents the sample size. The residual or error terms ε_i, $i = 1, ... n$ are assumed to be independent random variables having a normal distribution with mean zero and constant variance σ^2. So the response variable y also has a normal distribution with expected value $E(y \mid x_1, x_2, \cdots, x_p) = \beta_0 + \beta_1 x_1 + \cdots + \beta_p x_p$ and variance σ^2.

The parameters of the model $\beta_k, k = 1, 2, \cdots, p$ are known as *regression coefficients*. They represent the expected change in the response variable associated with a unit change in the corresponding explanatory variable *conditional* on the remaining explanatory variables being held constant.

(NB: The 'linear' in multiple linear regression and in simple linear regression applies to the regression parameters, *not* to the response or explanatory variables. Consequently, models in which, for example, the logarithm of a response variable is modelled in terms of quadratic functions of some of the explanatory variables would be included in this class of models.)

The multiple regression model can be written most conveniently for all n individuals by using appropriate matrices and vectors as

$$\mathbf{y} = \mathbf{X}\boldsymbol{\beta} + \boldsymbol{\varepsilon}$$

where $\mathbf{y}' = [y_1, y_2, \cdots, y_n], \boldsymbol{\beta}' = [\beta_0, \beta_1, \cdots, \beta_p], \boldsymbol{\varepsilon}' = [\varepsilon_1, \varepsilon_2, \cdots, \varepsilon_n]$ and

$$\mathbf{X} = \begin{bmatrix} 1 & x_{11} & x_{12} & \cdots & x_{1p} \\ 1 & x_{21} & x_{22} & \cdots & x_{2p} \\ \vdots & \vdots & \vdots & \vdots & \vdots \\ 1 & x_{n1} & x_{n2} & \cdots & x_{np} \end{bmatrix}$$

Each row in $\mathbf{X}$ (sometimes known as the *design matrix*) represents the values of the explanatory variables for one of the individuals in the sample, with the addition of unity which takes care of the parameter β_0 needed in the model for each individual in the sample. Assuming that $\mathbf{X}'\mathbf{X}$ is nonsingular (i.e., can be inverted), then the least squares estimator of the parameter vector $\boldsymbol{\beta}$ is

$$\hat{\boldsymbol{\beta}} = (\mathbf{X}'\mathbf{X})^{-1}\mathbf{X}'\mathbf{y}$$

This estimator $\hat{\boldsymbol{\beta}}$ has the following properties

$$E(\hat{\boldsymbol{\beta}}) = \boldsymbol{\beta}, \operatorname{cov}(\hat{\boldsymbol{\beta}}) = \sigma^2 (\mathbf{X}'\mathbf{X})^{-1}$$

The diagonal elements of the matrix cov $(\hat{\boldsymbol{\beta}})$ give the variances of the $\hat{\beta}_j$, whereas the off-diagonal elements give the covariances between pairs $\hat{\beta}_j$, $\hat{\beta}_k$. The square roots of the diagonal elements of the matrix are thus the standard errors of the $\hat{\beta}_j$.

The fit of the regression model can be partially assessed, at least, by using the analysis of variance table shown below which partitions the total sum of squares of the response variable into a part due to regression on the explanatory variables and a part due to the errors in the model. In this table $\hat{y}_i$ is the predicted value of the response variable for the ith individual $\left(\hat{y}_i = \hat{\beta}_0 + \hat{\beta}_1 x_{ij} + \cdots + \hat{\beta}_p x_{ip}\right)$ and $\bar{y}$ is the mean value of the response variable.

SOURCE OF VARIATION	SUM OF SQUARES (SS)	DEGREES OF FREEDOM (DF)	MEAN SQUARE
Regression	$\sum_{i=1}^{n}(\hat{y}_i - \bar{y})^2$	p	MSR = SS/df
Residual	$\sum_{i=1}^{n}(y_i - \hat{y}_i)^2$	$n - p - 1$	MSE = SS/df
Total	$\sum_{i=1}^{n}(y_i - \bar{y})^2$	$n - 1$	

The mean square ratio MSR/MSE provides an F-test of the null hypothesis that the regression coefficients of all p explanatory variables take the value zero, i.e.,

$$H_0: \beta_1 = \beta_2 = \cdots = \beta_p = 0$$

Under H_0, the mean square ratio has an F-distribution with $p, n - p - 1$ degrees of freedom. (Testing this very general hypothesis is usually of limited interest.) An estimate of σ^2 is provided by s^2 given by

$$s^2 = \frac{1}{n-p-1}\sum_{i=1}^{n}(y_i - \hat{y}_i)^2$$

The correlation between the observed values y_i and the predicted values $\hat{y}_i$, R, is known as the *multiple correlation coefficient*. The value of R^2 gives the proportion of variance of the response variable accounted for by the explanatory variables.

EXAMPLE 6.2

The data in the table below give the values of the following variables for ten cities in the United States:

y: Sulphur dioxide content of air in micrograms per cubic metre (SO_2)

x_1: Average annual temperature (F)

x_2: Number of manufacturing enterprises employing 20 or more workers

x_3: Population size in thousands

CITY	y	x_1	x_2	x_3
Phoenix	10	70.3	213	582
Little Rock	13	61.0	91	132
San Francisco	12	56.7	453	716
Denver	17	51.9	454	515
Hartford	56	49.1	412	158
Wilmington	36	54.0	80	80
Washington	29	57.3	434	757
Jacksonville	14	68.4	136	529
Miami	10	75.5	207	335
Atlanta	24	61.5	368	497

Fit a multiple linear model to the data.

SOLUTION

PARAMETER	ESTIMATE	ESTIMATED STANDARD ERROR
$\hat{\beta}_0$	71.33	42.39
$\hat{\beta}_1$	0.73	0.65
$\hat{\beta}_2$	0.034	0.043
$\hat{\beta}_3$	− 0.034	0.024

$s^2 = 11.24$ on 6 degrees of freedom.

Multiple $R^2 = 0.6138$, F-statistic = 3.178 on 3 and 6 DF, p-value = 0.1061.

There is no evidence that any of the three regression coefficients differ from zero (with only 10 observations the power of the F-test is, of course, low).

6.3 SELECTING A PARSIMONIOUS MODEL

Parsimonious model: The simplest model that adequately describes the data. In multiple regression a model with a subset of the explanatory variables that predicts the response variable almost as well as the model containing all the explanatory variables.

t-test for assessing the hypothesis that a regression coefficient is zero: The t-statistic is $t = \hat{\beta}_i / \text{se}(\hat{\beta}_i)$ tested against a t-distribution with $n - p - 1$ df. But the test is not always helpful in searching for variables that can be removed from a model because it is conditional on the variables in the current model. Removing a variable will alter the estimated regression coefficients of the variables left in the model and the standard errors of these coefficients.

All possible subsets regression: A procedure that looks at all $2^p - 1$ regression models when there are p explanatory variables and compares each of them using some numerical criterion that indicates which models are 'best'. The most common criterion is Mallows' C_p statistic:

$$ C_p = \frac{\text{RSS}_p}{s^2} - (n - 2p) $$

where RSS_p is the residual sum of squares from a multiple linear regression model with a particular set of $p - 1$ explanatory variables plus an intercept and s^2 is an estimate of σ^2 usually obtained from the model including *all* the available explanatory variables. (NB: We are using p here for the number of variables in the putative model, not for the number of available explanatory variables.) It can be shown that C_p is an unbiased estimate of the mean square error, $E[\Sigma \hat{y}_i - E(y_i)]^2 / n$, of the model's fitted values as estimates of the true expectations of the observations. 'Low' values of C_p are those that indicate the best models to consider.

EXAMPLE 6.3
The data below show the weight and various physical measurements for 22 male subjects aged 16 to 30 years. Weight is in kilograms and all other measurements are in centimetres.
   Fit a multiple linear regression model to the data and use all subsets regression to select a parsimonious model.

	WEIGHT	FORE	BICEP	CHEST	NECK	SHOULDER	WAIST	HEIGHT	CALF	THIGH	HEAD
1	77.0	28.5	33.5	100.0	38.5	114.0	85.0	178.0	37.5	53.0	58.0
2	85.5	29.5	36.5	107.0	39.0	119.0	90.5	187.0	40.0	52.0	59.0
3	63.0	25.0	31.0	94.0	36.5	102.0	80.5	175.0	33.0	49.0	57.0
4	80.5	28.5	34.0	104.0	39.0	114.0	91.5	183.0	38.0	50.0	60.0
5	79.5	28.5	36.5	107.0	39.0	114.0	92.0	174.0	40.0	53.0	59.0
6	94.0	30.5	38.0	112.0	39.0	121.0	101.0	180.0	39.5	57.5	59.0
7	66.0	26.5	29.0	93.0	35.0	105.0	76.0	177.5	38.5	50.0	58.5
8	69.0	27.0	31.0	95.0	37.0	108.0	84.0	182.5	36.0	49.0	60.0
9	65.0	26.5	29.0	93.0	35.0	112.0	74.0	178.5	34.0	47.0	55.5
10	58.0	26.5	31.0	96.0	35.0	103.0	76.0	168.5	35.0	46.0	58.0
11	69.5	28.5	37.0	109.5	39.0	118.0	80.0	170.0	38.0	50.0	58.5
12	73.0	27.5	33.0	102.0	38.5	113.0	86.0	180.0	36.0	49.0	59.0
13	74.0	29.5	36.0	101.0	38.5	115.5	82.0	186.5	38.0	49.0	60.0
14	68.0	25.0	30.0	98.5	37.0	108.0	82.0	188.0	37.0	49.5	57.0
15	80.0	29.5	36.0	103.0	40.0	117.0	95.5	173.0	37.0	52.5	58.0
16	66.0	26.5	32.5	89.0	35.0	104.5	81.0	171.0	38.0	48.0	56.5
17	54.5	24.0	30.0	92.5	35.5	102.0	76.0	169.0	32.0	42.0	57.0
18	64.0	25.5	28.5	87.5	35.0	109.0	84.0	181.0	35.5	42.0	58.0
19	84.0	30.0	34.5	99.0	40.5	119.0	88.0	188.0	39.0	50.5	56.0
20	73.0	28.0	34.5	97.0	37.0	104.0	82.0	173.0	38.0	49.0	58.0
21	89.0	29.0	35.5	106.0	39.0	118.0	96.0	179.0	39.5	51.0	58.5
22	94.0	31.0	33.5	106.0	39.0	120.0	99.5	184.0	42.0	55.0	57.0

SOLUTION

Analysis of variance table:

SOURCE	SS	DF	MS	MSR	p-VALUE
Regression	2243.01	6	373.84	19.95	<0.001
Error	281.13	15	18.74		

Parameter estimates:

VARIABLE	PARAMETER ESTIMATE	STANDARD ERROR	ESTIMATE/ SE
Intercept	−69.52	29.04	−2.39
Forearm	1.78	0.85	2.09
Bicep	0.15	0.49	0.32
Chest	0.19	0.23	0.84
Neck	−0.48	0.72	−0.67
Waist	0.66	0.12	5.68
Height	0.32	0.13	2.44
Shoulder	−0.03	0.24	−0.12
Calf	0.45	0.41	1.08
Thigh	0.30	0.31	1.00
Head	−0.92	0.52	−1.768

$R^2 = 0.98$; the 10 explanatory variables account for 98% of the variation in weight.

Mallows' C_p statistic for a subset of models:

NUMBER OF VARIABLES IN MODEL	C_p	R^2	VARIABLES IN MODEL
5	4.14	0.97	Forearm, Waist, Height, Thigh, Head
6	4.38	0.97	Forearm, Waist, Height, Calf, Thigh, Head
4	4.44	0.97	Forearm, Waist, Height, Thigh
6	4.81	0.97	Forearm, Waist, Height, Thigh
5	4.82	0.97	Forearm, Waist, Height, Calf, Thigh
5	5.35	0.97	Forearm, Waist, Height, Calf, Head
6	5.50	0.97	Forearm, Chest, Waist, Height, Thigh, Head

The first two models in the table above have the lowest C_p values and both account for almost as much variation in weight as the model that includes all 10 explanatory available variables.

Stepwise selection methods: Methods for selecting variables to include in a multiple linear model that use a suitable numerical criterion approach to selection. There are three possibilities:

Forward selection: Begins with an initial model that contains only an intercept term and successively adds explanatory variables to the model until the pool of candidate variables remaining contains no variables that if added to the current model would contribute information that is statistically important for predicting the response.

Backward elimination: Begins with an initial model that contains all the available explanatory variables and successively removes variables until no variables amongst those remaining in the model can be eliminated without adversely affecting the ability of the model to predict the response.

Stepwise method: Combines elements of both forward selection and backward elimination. The initial model is one containing only an intercept. Subsequent cycles involve first the possible addition of a variable to the current model, followed by the possible elimination of one of the variables included earlier if the presence of new variables has made its contribution to the prediction of the response no longer important.

EXAMPLE 6.4
Use backward elimination with a suitable criterion for removing variables to find a parsimonious model for the physical measurements data.

SOLUTION
Here we shall use *Akaike's information criterion* (AIC) to decide whether a variable can be removed from the current candidate model:

$$AIC = -2L_{max} + 2m$$

where L_{max} is the maximized log-likelihood of the model and m is the number of parameters in the model. AIC takes into account both the statistical goodness of fit and the number of parameters needed to achieve this degree of fit. In a series of competing models lower values of the AIC indicate the preferred model.

The results of the backward elimination approach are

Start: AIC = 43.15

Explanatory variables in the model: forearm, bicep, chest, neck, waist, height, shoulder, calf, thigh and head.

Step 1: Removing one explanatory variable at a time and leaving the remaining five in the model

Remove shoulder: Model AIC = 41.18
Remove bicep: Model AIC = 41.35
Remove neck: Model AIC = 42.02
Remove chest: Model AIC = 42.501
Remove thigh: Model AIC = 42.97
Remove calf: Model AIC = 43.34
Remove head: Model AIC = 46.65
Remove forearm: Model AIC = 48.47
Remove height: Model AIC = 50.65
Remove waist: Model AIC = 71.26

Remove shoulder, leaving a model with nine explanatory variables and a lower AIC than the original ten explanatory variable model.
AIC = 41.18, variables in model, neck, waist, chest, height, forearm, bicep, calf, thigh, head.

Step 2: Removing one explanatory variable at a time and leaving the other eight in the model

Remove bicep: Model AIC = 39.40
Remove neck: Model AIC = 40.06
Remove chest: Model AIC = 40.90
Remove calf: Model AIC = 41.50
Remove thigh: Model AIC = 41.54
Remove head: Model AIC = 45.33
Remove forearm: Model AIC = 49.31
Remove height: Model AIC = 50.06
Remove waist: Model AIC = 69.87

Remove bicep, leaving a model with eight explanatory variables and a lower AIC than the nine explanatory variable model.
AIC = 39.40, variables in model, waist, chest, height, forearm, neck, calf, thigh, head.

Step 3: Removing one explanatory variable at a time and leaving the other seven in the model

Remove neck: Model AIC = 38.06
Remove thigh: Model AIC = 39.65
Remove chest: Model AIC = 39.74
Remove calf: Model AIC = 39.98
Remove head: Model AIC = 43.37
Remove forearm: Model AIC = 49.34
Remove height: Model AIC = 50.43
Remove waist: Model AIC = 67.92

Remove neck, leaving a model with seven explanatory variables and a lower AIC than the eight explanatory variable model.

AIC = 38.06, variables in model, waist, chest, height, forearm, calf, thigh.

Remove chest: Model AIC = 37.74
Remove thigh: Model AIC = 38.50
Remove calf: Model AIC = 39.65
Remove head: Model AIC = 41.58
Remove forearm: Model AIC = 48.09
Remove height: Model AIC = 48.70
Remove waist: Model AIC = 66.39

Remove chest leaving a model with six explanatory variables and a lower AIC than the seven explanatory variable model

There is no five-variable model with a lower AIC value than that for the six-variable model containing variables including waist, height, forearm, calf, thigh and head; this is the model chosen.

6.4 REGRESSION DIAGNOSTICS

Regression diagnostics: Methods for determining whether or not the assumptions made in fitting a multiple linear regression mode are valid.

The hat matrix: The matrix, $\mathbf{H} = \mathbf{X}(\mathbf{X'X})^{-1}\mathbf{X'}$. The predicted values of the response are $\hat{\mathbf{y}} = \mathbf{Hy}$ so that $\mathbf{H}$ 'puts the hats' on $\mathbf{y}$. The diagonal elements of $\mathbf{H}, h_{ii}, i = 1, 2, \ldots, n$ are such that $0 \leq h_{ii} \leq 1$ and have an average value of p/n. Observations with high values of h_{ii} are said to have high *leverage*; such observations have the most effect on the estimation of model parameters. A plot of h_{ii} against i may help to identify observations that have undue influence when fitting the model.

Residual: The difference between an observed value of the response variable and its predicted value from the fitted model, $r_i = y_i - \hat{y}_i$.

Plotting residuals: The following plots involving the residuals are useful when assessing model assumptions.

- Residuals versus fitted values. If the fitted model is appropriate, the plotted points should lie in an approximately horizontal band across the plot. Departures from this appearance may indicate that the functional form of the assumed model is incorrect or, alternatively, that there is non-constant variance.

- Residuals versus explanatory variables. Systematic patterns in these plots can indicate violations of the constant variance assumption or an inappropriate model form.
- Normal probability plot of the residuals. The plot checks the normal distribution assumptions on which all statistical inference procedures are based.

Figure 6.2 shows some idealized plots that indicate particular points about models: Figure 6.2(a) is what is looked for to confirm that the fitted model meets the assumptions of the regression model; Figure 6.2(b) suggests that the assumption of constant variance is not justified so a transformation of the response variable before fitting might be a sensible option to consider; and

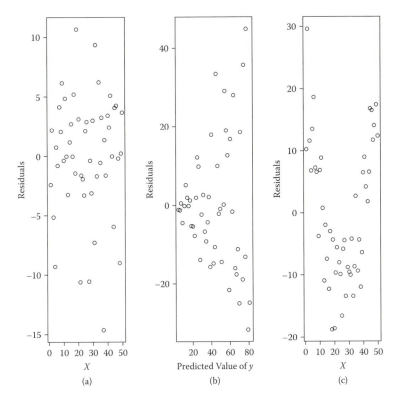

FIGURE 6.2   Idealized residual plots.

Figure 6.2(c) implies that the model requires a quadratic term in the explanatory variables used in the plot.

Modified residuals: The 'raw' residuals are not independent of one another nor do they have the same variance—the variance of $r_i = y_i - \hat{y}_i$ is $\sigma^2(1 - h_{ii})$. Two alternative residuals are the *standardized residual* and the *deletion residual*:

$$r_i^{\text{std}} = \frac{r_i}{\sqrt{s^2(1 - h_{ii})}}$$

where s^2 is the residual mean square estimate of σ^2.

$$r_i^{\text{del}} = \frac{r_i}{\sqrt{s_{(-i)}^2(1 - h_{ii})}}$$

where $s_{(-i)}^2$ is the residual mean square found from the $n - 1$ observations remaining after deletion of observation i.

Cook's distance: Another regression diagnostic that measures the influence of the ith observation on the estimation of all the parameters in the model. Values greater than one suggest that an observation has undue influence on the estimation process. The diagnostic is defined as

$$D_i = \frac{r_i h_{ii}}{ps^2(1 - h_{ii})}$$

DFBETA: A statistic that measures the influence of a particular observation, i, on a specific estimated regression coefficient, that for variable j. Given by the standardized change in the regression coefficient when a particular observation is deleted from the analysis,

$$\text{DFBETA}_{j(-i)} = \frac{\hat{\beta}_j - \hat{\beta}_{j(-i)}}{s_{(-i)}\sqrt{c_j}}$$

where C_j is the $(j + 1)$th diagonal element of $(\mathbf{X'X})^{-1}$.

DFFITS: A statistic that measures the influence of an observation on the predicted response value. Defined as

$$\text{DFFITS}_i = \frac{\hat{y}_i - \hat{y}_{i(-i)}}{s_{(-i)}\sqrt{h_i}}$$

Some examples of a number of different types of diagnostic plots for the model selected by the backward selection method applied to the physical measurements data are shown in Figures 6.3, 6.4 and 6.5.

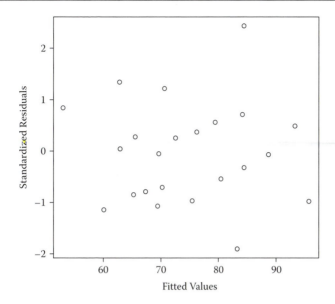

FIGURE 6.3   Standardized residuals against fitted values.

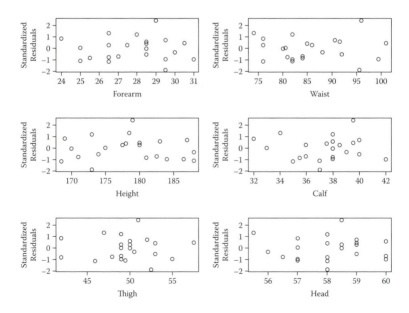

FIGURE 6.4   Standardized residuals against each explanatory variable.

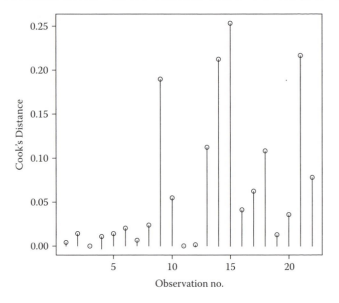

FIGURE 6.5 Index plot of Cook's distances.

6.5 ANALYSIS OF VARIANCE AS REGRESSION

Consider a one-way analysis of variance where there are three groups and two observations per group. The model is

$$y_{ij} = \mu + \alpha_i + \varepsilon_{ij}$$

This can be written as

$$y = X\beta + \varepsilon$$

where $y' = [y_{11}, y_{12}, y_{21}, y_{22}, y_{31}, y_{32}]$, $\hat{\beta} = [\mu, \alpha_1, \alpha_2]$, $\varepsilon' = [\varepsilon_{11}, \varepsilon_{12}, \varepsilon_{21}, \varepsilon_{22}, \varepsilon_{31}, \varepsilon_{32}]$ and

$$\mathbf{X} = \begin{bmatrix} 1 & 1 & 0 & 0 \\ 1 & 1 & 0 & 0 \\ 1 & 0 & 1 & 0 \\ 1 & 0 & 1 & 0 \\ 1 & 0 & 0 & 1 \\ 1 & 0 & 0 & 1 \end{bmatrix}$$

This is the same as the model of multiple linear regression but here the matrix $\mathbf{X'X}$ is singular and cannot be inverted because of the linear dependency in the columns of $\mathbf{X}$—the sum of columns 2, 3 and 4 equal column 1. The model is *over parameterised*. But if we introduce the constraint $\sum_{i=1}^{3} \alpha_i = 0$, i.e., $\alpha_3 = -\alpha_1 - \alpha_2$ the model can be rewritten as

$$\mathbf{y} = \mathbf{X}\hat{\boldsymbol{\beta}} + \boldsymbol{\varepsilon}$$

where now $\mathbf{y'} = [y_{11}, y_{12}, y_{21}, y_{22}, y_{31}, y_{32}]$, $\hat{\boldsymbol{\beta}} = [\mu, \alpha_1, \alpha_2]$, $\boldsymbol{\varepsilon'} = [\varepsilon_{11}, \varepsilon_{12}, \varepsilon_{21}, \varepsilon_{22}, \varepsilon_{31}, \varepsilon_{32}]$ and

$$\mathbf{X} = \begin{bmatrix} 1 & 1 & 0 \\ 1 & 1 & 0 \\ 1 & 0 & 1 \\ 1 & 0 & 1 \\ 1 & -1 & -1 \\ 1 & -1 & -1 \end{bmatrix}$$

Now the matrix $\mathbf{X'X}$ *is* invertible and the model can be dealt with as a multiple linear regression model. Analysis of variance and multiple linear regression are essentially equivalent procedures.

Two-way analysis of variance as multiple regression: Consider a 2×2 factorial design with factors A and B both at two levels, A1 and A2, and B1 and B2. The usual analysis of variance model for such a design is

$$y_{ijk} = \mu + \alpha_i + \beta_j + \gamma_{ij} + \varepsilon_{ijk}$$

where y_{ijk} represents the kth observation in the ijth cell of the design, α_i represents the effect of the ith level of factor A, β_j represent the effect of the jth level of factor B, γ_{ij} represent the interaction of A and B and as always ε_{ijk} represent random error terms with the usual distributional assumptions. The usual constraints on the parameters to deal with over parameterization in this case are

$$\sum_{i=1}^{2}\alpha_i = 0, \ \sum_{j=1}^{2}\beta_j = 0, \ \sum_{i=1}^{2}\gamma_{ij} = \sum_{j=1}^{2}\gamma_{ij} = 0$$

These constraints imply that the parameters in the model are such that

$$\alpha_1 = -\alpha_2, \ \beta_1 = -\beta_2, \ \gamma_{1j} = -\gamma_{2j}, \ \gamma_{i1} = -\gamma_{i2}$$

The last two equations imply that

$$\gamma_{12} = -\gamma_{11}, \ \gamma_{21} = -\gamma_{11}, \ \gamma_{22} = \gamma_{11}$$

showing that there is only a single interaction parameter. The model for the observations in each of the four cells of the design can now be written explicitly as follows:

	A1	A2
B1	$\mu + \alpha_1 + \beta_1 + \gamma_{11}$	$\mu - \alpha_1 + \beta_1 - \gamma_{11}$
B2	$\mu + \alpha_1 - \beta_1 - \gamma_{11}$	$\mu - \alpha_1 - \beta_1 + \gamma_{11}$

Now we define two variables as follows:

$x_1 = 1$ if first level of A, $x_1 = -1$ if second level of A
$x_2 = 1$ if first level of B, $x_2 = -1$ if second level of B

The original ANOVA model for the design can now be written as

$$y_{ijk} = \mu + \alpha_1 x_1 + \beta_1 x_2 + \gamma_{11} x_3 + \varepsilon_{ijk} \text{ where } x_3 = x_1 \times x_2$$

We can now recognise this as a multiple linear regression model with three explanatory variables and we can fit it in the usual way. Here the fitting process can be used to illustrate the difference in analysing a *balanced* 2 x 2 design (equal number of observations per cell) and *unbalanced* design (unequal number of observations per cell). To begin we will apply the multiple regression model to the data in Table 6.1

So for fitting the multiple regression model all observations in cell A1,B1 have $x_1 = 1$ and $x_2 = 1$, all observations in cell A1,B2 have $x_1 = 1$, $x_2 = -1$ and so

TABLE 6.1   A balanced
2 x 2 data set

	A1	A2
B1	23	22
	25	23
	27	21
	29	21
B2	26	37
	32	38
	30	40
	31	35

on for the remaining observations in Table 6.1. To begin we will fit the model with the single explanatory variable, x_1, to give the following results:

SOURCE	SUM OF SQUARES	DF	MEAN SQUARE
Regression	12.25	1	12.25
Residual	580.75	14	41.48

and $\hat{\mu} = 28.75$, $\hat{\alpha}_1 = -0.875$

The regression sum of squares of 12.25 is what would be the between levels of A sum of squares in an analysis of variance table. Now fit the regression with x_1 and x_2 as explanatory variables to give

SOURCE	SUM OF SQUARES	DF	MEAN SQUARE
Regression	392.50	2	196.25
Residual	200.50	13	15.42

and $\hat{\mu} = 28.75$, $\hat{\alpha}_1 = -0.875$, $\hat{\beta}_1 = -4.875$.

The difference between the regression sums of squares for the two-variable model and the one-variable model gives the sum of squares for factor B that would be obtained in an analysis of variance. Lastly we can fit a model with three explanatory variables to give

SOURCE	SUM OF SQUARES	DF	MEAN SQUARE
Regression	536.50	3	178.83
Residual	56.50	12	4.71

and $\hat{\mu} = 28.75$, $\hat{\alpha}_1 = -0.875$, $\hat{\beta}_1 = -4.875$, $\gamma_{11} = 3.000$.

The difference between the regression sums of squares for the three-variable model and the two-variable model gives the sum of squares for the A × B interaction that would be obtained in an analysis of variance. The residual sum of squares in the final step corresponds to the error sum of squares in the usual ANOVA table. Note that unlike the estimated regression coefficients in the examples considered earlier, the estimated regression coefficients for the balanced 2 × 2 design do *not* change as extra explanatory variables are introduced into the regression model. The factors in a balanced design are *independent*— a more technical term is that they are *orthogonal*. When the explanatory variables are orthogonal, adding variables to the regression model in a different order than the one used above will alter nothing; the corresponding sums of squares and regression coefficient estimates will be the same.

Now we will repeat the above but with an unbalanced data set (see Table 6.2).

TABLE 6.2 Unbalanced 2 × 2 data set

	A1	A2
B1	23	22
	25	23
	27	21
	29	21
	30	19
	27	23
	23	17
	25	
B2	26	37
	32	38
	30	40
	31	35
		39
		35
		38
		41
		32
		36
		40
		41
		38

Again we will fit regression models first with only x_1, then x_1 and x_2 and lastly with x_1, x_2 and x_3.

Results for x_1 model:

SOURCE	SUM OF SQUARES	DF	MEAN SQUARE
Regression	149.63	1	149.63
Residual	1505.87	30	50.19

and $\hat{\mu} = 29.567$, $\hat{\alpha}_1 = -2.233$.

The regression sum of squares gives the sum of squares for factor A.

Results for x_1 and x_2 model:

SOURCE	SUM OF SQUARES	DF	MEAN SQUARE
Regression	1180.86	2	590.42
Residual	476.55	29	16.37

and $\hat{\mu} = 29.667$, $\hat{\alpha}_1 = -0.341$, $\hat{\beta}_1 = -5.997$.

The difference in the regression sums of squares for the two-variable and one-variable models gives the sum of squares due to factor B, *conditional* on A already being in the model.

Results for x_1, x_2 and x_3 model:

SOURCE	SUM OF SQUARES	DF	MEAN SQUARE
Regression	1474.25	3	491.42
Residual	181.25	28	6.47

and $\hat{\mu} = 28.606$, $\hat{\alpha}_1 = -0.667$, $\hat{\beta}_1 = -5.115$, $\hat{\gamma}_{11} = 3.302$.

The difference in the regression sums of squares for the three-variable and two-variable models gives the sum of squares due to the A × B interaction, *conditional* on A and B being in the model.

For an unbalanced design the factors are no longer orthogonal and so the estimated regression parameters change as further variables are added to the model, and the sums of squares for each term in the model are now conditional on what has entered the model before them. If variable x_2 was entered before x_1 then the results would differ from those given above.

So using the regression approach clearly demonstrates why there is a difference in analysing a balanced design (not just a 2 × 2 design as in the example) from analysing an unbalanced design. In the latter, *order* of entering effects is important.

6.6 SUMMARY

Simple linear regression: Model $y_1 = \beta_0 + \beta_1 x_i + \varepsilon_i$, $= 1 \ldots n$, Least squares estimates of parameters,

$$\hat{\beta}_0 = \bar{y} - \hat{\beta}_1 \bar{x}, \ \hat{\beta}_1 = \frac{\sum_{i=1}^{n}(y_i - \bar{y})(x_i - \bar{x})}{\sum_{i=1}^{n}(x_i - \bar{x})^2}$$

Multiple linear regression: Model $\mathbf{y} = \mathbf{X}\boldsymbol{\beta} + \boldsymbol{\varepsilon}$, Least squares estimates, $\hat{\boldsymbol{\beta}} = (\mathbf{X}'\mathbf{X})^{-1}\mathbf{X}'\mathbf{Y}$

The hat matrix: The matrix, $\mathbf{H} = \mathbf{X}(\mathbf{X}'\mathbf{X})^{-1}\mathbf{X}'$.

Standardized residual: $r_i^{\text{std}} = \dfrac{r_i}{\sqrt{s^2(1 - h_{ii})}}$

Deletion residual: $r_i^{\text{del}} = \dfrac{r_i}{\sqrt{s_{(-i)}^2(1 - h_{ii})}}$

Cook's distance: $D_i = \dfrac{r_i h_{ii}}{\sqrt{ps^2(1 - h_{ii})}}$

DFBETA$_{j(-i)} = \dfrac{\hat{\beta}_j - \hat{\beta}_{j(-i)}}{s_{(-i)}\sqrt{c_j}}$

DFFITS$_i = \dfrac{\hat{y}_i - \hat{y}_{i(-i)}}{s_{(-i)}\sqrt{h_i}}$

SUGGESTED READING

Allison, PD (1998) *Multiple Regression: A Primer*, Pine Forge Press, Thousand Oaks, CA.

Chatterjee, S and Hadi, AS (2006) *Regression Analysis by Example*, 3rd edition, Wiley, New York.

Draper, NR and Smith, H (1998) *Applied Regression Analysis*, 3rd edition, Wiley, New York

Logistic Regression and the Generalized Linear Model

7

7.1 ODDS AND ODDS RATIOS

Odds: If p is the probability of an event, $p/(1 - p)$ is the odds of the event.
The odds ratio: The ratio of the odds of an event in two different groups.
Estimating the odds ratio: Consider the general 2×2 contingency table give by

VARIABLE 1	VARIABLE 2	
	CATEGORY 1	CATEGORY 2
Category 1	a	b
Category 2	c	d

The odds ratio in the population will be denoted by ψ and it can be estimated from the observed frequencies in the table as the ratio of the estimated odds of category 1 to category 2 of variable 1 in category 1 of variable 2 to the corresponding estimated odds in category 2 of variable 2, i.e.,

$$\hat{\psi} = a/c \div b/d = \frac{ad}{bc}$$

Estimator of the variance of log($\hat{\psi}$): $\hat{var}(\log \hat{\psi}) = 1/a + 1/b + 1/c + 1/d$
Approximate 95% confidence interval for $\log(\psi)$:

$$\log(\hat{\psi}) \pm 1.96 \times \sqrt{\hat{var}(\log \hat{\psi})}$$

If the limits of the confidence interval for $\log(\psi)$ obtained in this way are ψ_L, ψ_U then the corresponding confidence for ψ is $\exp(\psi_L)$ and $\exp(\psi_U)$.

EXAMPLE 7.1

The following is a 2 × 2 contingency table of gender against possible psychiatric 'caseness' in a sample of people.

	CASE	NONCASE
Male	25	79
Female	43	131

Calculate a confidence interval for the population odds ratio.

SOLUTION

The estimated odds of 'caseness' for women can be calculated directly from the frequencies in the 2 × 2 table as 43/131 = 0.328; the corresponding odds for men is 25/79 = 0.316 and the estimated odds ratio is 0.316/0.328 = 0.963. When the two variables forming the contingency table are independent, the odds ratio in the population will be 1. Log($\hat{\psi}$) = −0.037 and the estimated variance of log($\hat{\psi}$) is 1/43 + 1/131 + 1/25 + 1/79 = 0.084, leading to a 95% confidence interval for $\log(\psi)$ of −[−0.036 − 1.96 × 0.290, − 0.036 + 1.96 × 0.290], i.e., [−0.604, 0.531]. Finally, the confidence interval for ψ itself is found as

[exp(−0.604), exp(0.531)], i.e., [0.546, 1.701].

As this interval contains the value 1, we can conclude that there is no evidence of an association between 'caseness' and gender.

7.2 LOGISTIC REGRESSION

Logistic regression: A form of regression analysis used when the response variable is binary or categorical with more than two categories. (NB: Here we will deal only with binary responses.)

The logistic regression model: Analogous to the multiple regression model, we *could* model the expected value of the response variable as a linear function of the p explanatory variables. The expected value of a binary response labelled 0 for 'failure' and 1 for 'success' is simply the probability of a success, π. The model would be

$$\pi = \beta_0 + \beta_1 x_1 + \beta_2 x_2 + \cdots \beta_p x_p$$

There are two problems with this model:

- The predicted value of the probability, π, must satisfy $0 \le \pi < 1$, whereas a linear predictor can yield values from minus infinity to plus infinity.
- The observed values of the response variable, y, conditional on the values of the explanatory variables, will not now follow a normal distribution with mean π but rather a *Bernoulli distribution* (essentially a single trial binomial distribution).

Consequently, some suitable transformation of π must be modelled. The transformation most often used is the *logit* function of the probability which is simply the log(odds), i.e., $\log\left(\dfrac{\pi}{1-\pi}\right)$ and this leads to the logistic regression model having the form

$$\mathrm{logit}(\pi) = \log\frac{\pi}{1-\pi} = \beta_0 + \beta_1 x_1 + \cdots \beta_p x_p$$

The logit transformation of π can take values ranging from $-\infty$ to $+\infty$ and thus overcomes the first problem associated with modelling π directly. In a logistic regression model the parameter β_i associated with explanatory variable x_i represents the expected change in the log odds when x_i is increased by one unit, conditional on the other explanatory variables remaining the same. Interpretation is simpler using $\exp(\beta_i)$ which corresponds to an odds ratio.

Alternative form of logistic regression model:

$$\pi = \frac{\exp(\beta_0 + \beta_1 x_1 + \cdots \beta_p x_p)}{1 + \exp(\beta_0 + \beta_1 x_1 + \cdots \beta_p x_p)}$$

Error term in logistic regression: For a binary response we can express an observed value, y, as $y = \pi(x_1, x_2, \ldots, x_p) + \varepsilon$ (note that here we have introduced a slight change in the nomenclature to remind us that the expected value of the response is dependent on the explanatory variables). The error term ε can assume only one of two possible values: if $y = 1$ then $\varepsilon = 1 - \pi(x_1, x_2, \ldots, x_p)$ with probability $\pi(x_1, x_2, \ldots, x_p)$, and if $y = 0$ then $\varepsilon = \pi(x_1, x_2, \ldots, x_p)$ with probability $1 - \pi(x_1, x_2, \ldots, x_p)$. Consequently, ε has a distribution with mean zero and variance equal to $\pi(x_1, x_2, \ldots, x_p)(1 - \pi(x_1, x_2, \ldots, x_p))$. The distribution of the response variable conditional on the explanatory variables is a Bernoulli distribution.

Estimation in the logistic regression model: The parameters of the logistic model, $\boldsymbol{\beta}' = [\beta_0, \beta_1, \ldots, \beta_p]$, can be estimated by maximum likelihood; the log of the likelihood is

$$l(\boldsymbol{\beta}; \boldsymbol{y}) = \sum_{i=1}^{n} \{y_i \log[\pi(\boldsymbol{\beta}' \mathbf{x}_i)] + (1 - y_i) \log[1 - \pi(\boldsymbol{\beta}' \mathbf{x}_i)]\}$$

where $\boldsymbol{y}' = [y_1, y_2, \ldots, y_n]$ are the n observed values of the dichotomous response variable and $\mathbf{x}_i' = [x_{i1}, x_{i2}, \ldots, x_{ip}]$ is the vector of values of the explanatory variables for the ith sample member. The log-likelihood is maximized using *Fisher's method of scoring.*

Deviance: A numerical criterion which can be used to assess the fit of logistic regression models and in particular to compare the fit of nested logistic regression models. The deviance is the ratio of the likelihood of the model of interest to the likelihood of the saturated model that fits the data perfectly. Differences in the deviance of two nested models can be used to choose between the models. The difference has an approximate chi-square distribution with degrees of freedom equal to the difference in the number of parameters of the two models.

EXAMPLE 7.2

The data below arise from a study of a psychiatric screening questionnaire called the *General Health Questionnaire* (GHQ). (Here the original binary responses of the individuals have been tabulated for convenience.)

PSYCHIATRIC CASENESS DATA

GHQ SCORE	SEX	NUMBER OF CASES	NUMBER OF NON-CASES
0	F	4	80
1	F	4	29
2	F	8	15
3	F	6	3
4	F	4	2
5	F	6	1
6	F	3	1
7	F	2	0
8	F	3	0
9	F	2	0
10	F	1	0
0	M	1	36
1	M	2	25
2	M	2	8
3	M	1	4
4	M	3	1
5	M	3	1
6	M	2	1
7	M	4	2
8	M	3	1
9	M	2	0
10	M	2	0

Fit a logistic model to the data to investigate how the probability of being judged a 'case' is related to gender and GHQ score.

SOLUTION

Model 1: Logistic regression model with GHQ as the single explanatory variable.

	ESTIMATE	STANDARD ERROR (SE)	ESTIMATE/SE
Intercept	−2.71	0.27	−9.95
GHQ	0.74	0.09	7.78

Null deviance 130.31 on 21 df; residual deviance 16.24 on 20 df.
Fitted model is

$$\log[\mathrm{pr(case)/pr(not\ case)}] = -2.71 + 0.74 \times \mathrm{GHQ\ score}$$

This equation can be rearranged to give the predicted probabilities for the fitted logistic regression model as

$$\mathrm{Pr(case)} = \frac{\exp(-2.71 + 0.74 \times \mathrm{GHQ\ score})}{[1 + \exp(-2.71 + 0.74 \times \mathrm{GHQ\ score})]}$$

For the individual with a GHQ score of 10 this model predicts the probability of the individual being judged a case as 0.99. The estimated odds ratio of being judged a case for a difference of one on the GHQ scale is $\exp(0.74) = 2.10$; the 95% confidence interval is $[\exp(0.74 - 1.96 \times 0.09), \exp(0.74 + 1.96 \times 0.09)] = [1.76, 2.50]$. The increase in the odds of being judged a case for a one unit increase in GHQ score is estimated to be between 76% and 150%. The null deviance is that for a model with no explanatory variables and the residual deviance is that of a model with GHQ as the single explanatory variable; the reduction in deviance is considerable and if tested as a chi-square with one degree of freedom would be very highly significant. Clearly GHQ is very predictive of the probability of being judged a case.

Figure 7.1 shows the probabilities of being judged a case as predicted by the model and as observed. (Also shown on the plot is the fitted linear model for the probability of caseness to demonstrate how with this model predicted probabilities greater than 1 can occur.)

Model 2: Logistic regression model with gender as the single explanatory variable.

	ESTIMATE	STANDARD ERROR (SE)	ESTIMATE/SE
Intercept	−1.11	0.18	−6.34
Sex (M = 1, F = 0)	−0.04	0.29	−0.13

Null deviance 130.31 on 21 df, residual deviance 130.29 on 20 df.

The reduction in deviance achieved by the model including only gender is very small; apparently gender has no significant effect on the probability of being judged a case.

Model 3: Logistic regression model with GHQ score and gender as explanatory variables.

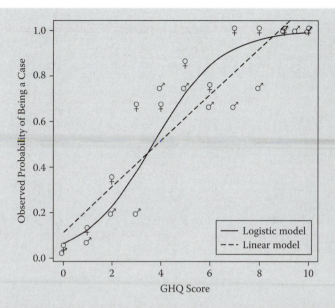

FIGURE 7.1 Plot of predicted probabilities of caseness from both linear and logistic regression models with GHQ score as the single explanatory variable and observed probabilities labelled by gender.

	ESTIMATE	STANDARD ERROR (SE)	ESTIMATE/SE
Intercept	−2.49	0.28	−8.85
Gender	−0.94	0.43	−2.16
GHQ	0.78	0.10	7.88

Null deviance 130.31 on 21 df, residual deviance 11.11 on 19 df.

Comparing the residual deviance of this model with that for model 1 we see a reduction of 5.13 with 1 df. Adding gender to the model including only GHQ produces a significant reduction in the deviance.

In this model the regression coefficient for gender shows that, conditional on the GHQ score, the log odds of caseness for men is −0.94 lower than for women, which gives an estimated odds ratio of exp(−0.94) = 0.39 with a 95% confidence interval of [0.167, 0.918]. For a given GHQ score the odds of a man being diagnosed as a case is between about 0.17 and 0.92 of the corresponding odds for a woman. But in model 2 where gender was the only explanatory variable, the gender effect was found to be not significant. We might ask why the difference? The reason is that the overall odds ratio is dominated by the large number of cases for the lower GHQ scores.

The estimated regression coefficient for GHQ conditional on gender is 0.779. This is very similar to the value for the model including only the GHQ score and so the interpretation of the conditional coefficient is very similar to that given previously.

The fitted model is shown in Figure 7.2.

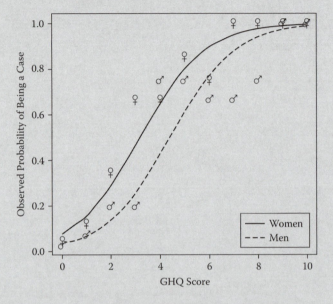

FIGURE 7.2 Plot of predicted probabilities of caseness from logistic regression model with gender and GHQ score as explanatory variables and observed probabilities labelled by gender.

Model 4: Logistic regression model with GHQ, gender and their interaction as explanatory variables.

	ESTIMATE	STANDARD ERROR (SE)	ESTIMATE/SE
Intercept	−2.77	0.36	−7.73
Gender	−0.26	0.61	−0.37
GHQ	0.94	0.16	6.00
Gender x GHQ	−0.30	0.20	−1.52

Residual deviance 8.77 on 18 df.

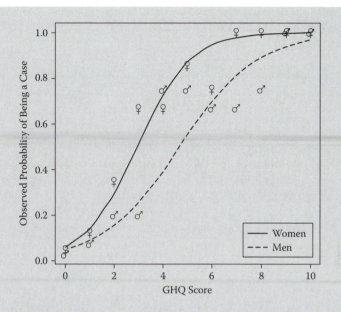

FIGURE 7.3 Plot of predicted probabilities of caseness from logistic regression model with gender, GHQ score and GHQ x gender as explanatory variables, and observed probabilities labelled by gender.

Comparing the residual deviance with that for model 3 we see that the addition of the interaction term has resulted in a decrease in deviance of 2.34 with 1 df. Testing this as a chi-square gives a p-value of 0.13; model 4 gives no significant improvement in fit over model 3. Model 3 is the most appropriate model for the data. Nevertheless, for interest Figure 7.3 shows the fitted interaction model.

7.3 GENERALIZED LINEAR MODEL

Generalized linear model (GLM): A unified framework for a wide range of seemingly disparate techniques of statistical modelling and inference that includes analysis of variance, analysis of covariance, multiple linear regression and logistic regression.

Components of a GLM: In a GLM a transformation of the expected value of the response variable is modelled as a linear function of the explanatory variables and the distribution of the response variable around its mean (often referred to as the *error distribution*) is generalized in a way that fits naturally with a particular transformation. The three essential components of a GLM are

(a) A linear predictor, η, formed from the explanatory variables

$$\eta = \beta_0 + \beta_1 x_1 + \beta_2 x_2 \cdots + \beta_p x_p = \boldsymbol{\beta}'\mathbf{x}$$

(b) A transformation of the mean, μ, of the response variable called the *link function*, $g(\mu)$. In a GLM it is $g(\mu)$ which is modelled by the linear predictor $g(\mu) = \eta$. The inverse of the link function is sometimes called the *mean function*: $g^{-1}(\eta) = \mu$.

Commonly used link functions and their inverses are

Link	$\eta = g(\mu)$	$\mu = g^{-1}(\eta)$
Identity	μ	η
Log	$\text{Log}(\mu)$	e^η
Square root	$\sqrt{\mu}$	η^2
Logit	$\log\dfrac{\mu}{1-\mu}$	$\dfrac{1}{1+e^{-\mu}}$

In multiple linear regression and analysis of variance, the link function is the identity function. Other link functions that are used include the log, logit, probit, inverse and power transformations, although the log and logit are those most commonly met in practice. The logit link, for example, is the basis of logistic regression.

(c) The distribution of the response variable given its mean μ is assumed to be a distribution from the *exponential family*; this has the form

$$f(y;\theta,\phi) = \exp\{(y\theta - b(\theta))/a(\phi) + c(y,\phi)\}$$

for some specific functions a, b and c and parameters θ and ϕ. For example, in linear regression, a normal distribution is assumed with mean μ and constant variance σ^2. This can be expressed via the exponential family as follows:

$$f(y;\theta,\phi)=\frac{1}{\sqrt{(2\pi\sigma^2)}}\exp\left\{-(y-\mu)^2/2\sigma^2\right\}$$

$$=\exp\left\{(y\mu-\mu^2/2)/\sigma^2-\frac{1}{2}(y^2/\sigma^2+\log(2\pi\sigma^2))\right\}$$

so that $\theta=\mu, b(\theta)=\theta^2/2, \phi=\sigma^2$ and $a(\phi)=\phi$. Other distributions in the exponential family include the binomial, Poisson, gamma, inverse Gaussian and exponential distributions. Particular link functions in GLMs are generally associated with particular error distributions, for example, the identity link with the Gaussian distribution, the logit with the binomial, and the log with the Poisson.

Variance function: A function $V(\mu)$ which captures how the variance of the response variable depends on its mean. The general form of the relationship is Var(response) $=\phi V(\mu)$ where ϕ is known as the *scale factor*. The choice of probability distribution determines the relationships between the variance of the response variable (conditional on the explanatory variables) and its mean. For a number of common distributions we have

Normal: $V(\mu)=1, \phi=\sigma^2$.
Binomial: $V(\mu)=\mu(1-\mu), \phi=1$.
Poisson: $V(\mu)=\mu, \phi=1$.

(NB: For both the binomial and the Poisson distributions the variance function is completely determined by the mean; there is no free parameter for the variance because the scale parameter is fixed at 1. As we shall see in the next section this may be too restrictive in some applications.)

Estimation for GLMs: The parameters in GLMs are estimated by maximizing the joint likelihood of the observed responses given the parameters of the model and the explanatory variables. This approach uses an *iterated least squares* (IWLS) algorithm that proceeds by forming a quadratic local approximation to the log-likelihood function which is then maximized by linear weighted least squares. A brief account follows.

The vector $\beta^{(m)}$ contains the estimates of the regression coefficients of the GLM on the mth iteration. From these estimates we calculate the current values of the linear predictor, the fitted values, the variance function, the *working response* and the weights for the ith observation:

Linear predictor: $\eta_i^{(m)}=\mathbf{x}_i'\beta^{(m)}$
Fitted values: $\mu_i^{(m)}=g^{-1}(\eta_i^{(m)})$

Variance function: $v_i^{(m)} = V(\mu_i^{(m)}) / \phi$

Working response: $z_i^{(m)} = \eta_i^{(m)} + (y_i - \mu_i^{(m)})\left(\dfrac{\partial \eta_i}{\partial \mu_i}\right)^{(m)}$

Weights: $w_i^{(m)} = 1/c_i v_i^{(m)}[(\partial \eta_i / \partial \mu_i)^{(m)}]^2$ where the c_i are fixed constants (for example, in the binomial family $c_i = n_i^{-1}$)

New estimates of the regression coefficients are obtained by minimizing the weighted sum of squares

$$WS = \sum_{i=1}^{n} w_i (z_i - \mathbf{x}_i'\boldsymbol{\beta})^2$$

Iterations start with some suitable starting values for the regression coefficients and continue until convergence at the maximum likelihood estimates.

EXAMPLE 7.3

The data below arise from a study of familial polyposis (FAP), an autosomal dominant genetic defect which predisposes those affected to develop large numbers of polyps in the colon which if untreated may develop into colon cancer. Patients with FAP were randomly assigned to receive an active drug treatment or a placebo. The response variable was the number of colonic polyps at 3 months after starting treatment. Additional covariates of interest were number of polyps before starting treatment, gender and age.

		FAP DATA		
SEX	TREATMENT	BASELINE COUNT OF POLYPS	AGE	NUMBER OF POLYPS AT 3 MONTHS
0	1	7	17	6
0	0	77	20	67
1	1	7	16	4
0	0	5	18	5
1	1	23	22	16
0	0	35	13	31
0	1	11	23	6
1	0	12	34	20

1	0	7	50	7
1	0	318	19	347
1	1	160	17	142
0	1	8	23	1
1	0	20	22	16
1	0	11	30	20
1	0	24	27	26
1	1	34	23	27
0	0	54	22	45
1	1	16	13	10
1	0	30	34	30
0	1	10	23	6
0	1	20	22	5
1	1	12	42	8

Sex: 0 = female, 1 = male
Treatment: 0 = placebo, 1 = active
Is there any evidence of a treatment effect?

SOLUTION

Here the response variable is a count. For such responses Poisson regression is suitable; this form of regression is a GLM with a log link function and a Poisson error distribution.

	ESTIMATE	STANDARD ERROR (SE)	ESTIMATE/SE
Intercept	3.36	0.19	17.86
Gender	0.28	0.11	2.53
Treatment	-0.32	0.10	-3.23
Baseline	0.01	0.0004	22.25
Age	-0.03	0.007	-3.62

The scale parameter is fixed at 1. The deviance of the fitted model is 186.73 on 17 df. The 95% confidence interval for treatment effect is [exp(-0.32 - 1.96 × 0.10), exp(-0.32 + 1.96 × 0.10)] = [0.60, 0.88]. Patients receiving the active treatment are estimated to have between 60% and 88% the number of polyps at 3 months as patients receiving the placebo. (NB: The value of the deviance divided by the degrees of freedom has implications for the appropriateness of the fitted model, as we shall see in the next section.)

7.4 VARIANCE FUNCTION AND OVERDISPERSION

Overdispersion: In some cases when fitting GLMs with binomial or Poisson error distributions the constraints on the variance of the response produced by the variance function being completely determined by the mean is too restrictive to fully account for the empirical variance of the data, a phenomenon known as overdispersion. Overdispersion can often be spotted by comparing the deviance with its degrees of freedom. For a well-fitting model the two quantities should be approximately equal. If the deviance is far greater than the degrees of freedom, overdispersion may be indicated. A simple approach to the problem is to allow the scale parameter to be estimated from the data rather than fixing it at unity. The estimate is the residual deviance divided by its degrees of freedom, Parameter estimates remain the same but parameter standard errors are increased by multiplying them by the square root of the estimated scale parameter.

For the polyps data in Example 7.3 there is evidence of overdispersion; the estimated scale parameter is 10.98 and applying the square root of this value to the standard errors found from fitting the Poisson regression model leads to the treatment effect being non-significant. A possible reason for the occurrence of overdispersion in these data is that polyps do not occur independently of one another but instead may 'cluster' together. If so, it would lead to the extra variation observed.

7.5 DIAGNOSTICS FOR GLMS

Deviance residuals: $r_i^D = \text{sign}(y_i - \hat{\mu})\sqrt{d_i}$ where d_i is the contribution of the ith subject to the deviance.

Pearson residuals: $r_i^P = \dfrac{\phi^{1/2}(y_i - \hat{\mu})}{\sqrt{V(\hat{\mu}_i)}}$

Both these residuals can be used for detecting observations not well fitted by the model. The deviance residuals are more commonly used because their distribution tends to be closer to normal than that of the Pearson residuals.

A normal probability plot of the Pearson residuals from fitting the Poisson regression model to the polyps data is shown in Figure 7.4.

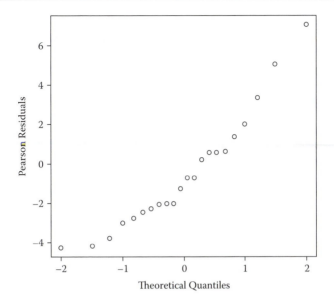

FIGURE 7.4 Normal probability plot of the Pearson residuals from the Poisson regression model fitted to the FAP data.

7.6 SUMMARY

Odds ratio estimator: $\hat{\psi} = \dfrac{ad}{bc}$

Variance of log (odds ratio estimator): $\hat{V}ar(\log \hat{\psi}) = 1/a + 1/b + 1/c + 1/d$

Approximate 95% Confidence interval for odds ratio:

$$\exp\left[\log(\hat{\psi}) \pm 1.96 \sqrt{\frac{1}{a} + \frac{1}{b} + \frac{1}{c} + \frac{1}{d}} \right]$$

Logistic regression model: $\log \text{it}(\pi) = \log \dfrac{\pi}{1-\pi} = \beta_0 + \beta_1 x_1 + \cdots \beta_p x_p$

Generalized linear model (GLM): A linear predictor, η

$$\eta = \beta_0 + \beta_1 x_1 + \beta_2 x_2 \cdots + \beta_p x_p = \boldsymbol{\beta}'\mathbf{x}$$

The *link function*, $g(\mu)$, and the model

$$g(\mu) = \eta$$

Variance function: A function $V(\mu)$ which captures how the variance of the response variable depends on its mean. The general form of the relationship is Var(response) $= \phi V(\mu)$ where ϕ is known as the *scale factor.*

Deviance residuals: $r_i^D = \text{sign}(y_i - \hat{\mu})\sqrt{d_i}$

Pearson residuals: $r_i^P = \dfrac{\phi^{1/2}(y_i - \hat{\mu})}{\sqrt{V(\hat{\mu}_i)}}$

SUGGESTED READING

Collett, D (2003) *Modelling Binary Data,* 2nd edition, Chapman & Hall/CRC, London.

Dobson, AJ and Barnett, A (2008) *An Introduction to Generalized Linear Models,* 3rd edition, Chapman & Hall/CRC, London.

Hosmer, D and Lemeshow, S (2000) *Applied Logistic Regression*, 2nd edition, Wiley, New York.

Survival Analysis

8

8.1 SURVIVAL DATA AND CENSORED OBSERVATIONS

Survival data: Observations of the time from an initiating event until the occurrence of some terminal event (such observations are also know as *time-to-event* data). Examples of initiating events are birth, start of treatment in a clinical trial, diagnosis of a particular condition and start of employment in a job. Examples of terminal events are death, relapse, remission and disability pension. If the end-point is death, the resulting data are literally survival times.

Special features of survival data:

1. Survival times are generally skewed; consequently, the assumption of normality is rarely justified.
2. Survival times are often *censored*, i.e., the terminal event is not observed for some individuals.

Types of censoring:

1. Right censoring: A patient who entered a study at time t_0 dies at time $t_0 + t$ but t is unknown because the individual is still alive when the study is stopped, at which point the individual has been in the study for time $t_0 + c$ where c is less than t and is known as a *censored survival time*. The right censored survival time is less than the actual, but unknown, survival time.
2. Left censoring: Occurs when the actual survival time is less than that observed. An example of such censoring is when patients are

seen only at intervals and the recurrence of a condition has happened in the interval. Less common than right censoring.

3. Interval censoring: Occurs when the terminal event occurs in an interval.

8.2 SURVIVOR FUNCTION, LOG-RANK TEST AND HAZARD FUNCTION

Survivor function, $S(t)$: Defined as the probability that the survival time, T, is greater than or equal to t, i.e.,

$$S(t) = \Pr(T > t)$$

Survivor function for the exponential distribution and the Weibull distribution: First the exponential with probability density function

$$f(t) = \lambda e^{-\lambda t}, 0 \le t < \infty$$

for which the survivor function is given by

$$S(t) = \int_t^\infty \lambda e^{-\lambda u} du = e^{-\lambda t}$$

Plots of the survivor functions of the exponential distribution for different values of λ are shown in Figure 8.1.

Now for the Weibull probability density function which is given by

$$f(t) = \lambda \gamma t^{\gamma-1} \exp(-\lambda t^\gamma), 0 \le t < \infty$$

The survivor function of the Weibull is given by

$$S(t) = \int_t^\infty \lambda \gamma u^{\gamma-1} \exp(-\lambda u^\gamma) du = \exp(-\lambda t^\gamma)$$

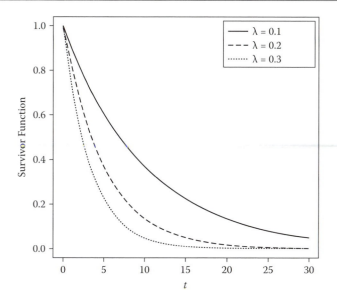

FIGURE 8.1 Survivor functions for a number of exponential distributions.

Plots of the survivor function of the Weibull distribution for different values of the scale parameter, λ, and the shape parameter, γ, are shown in Figure 8.2. **Estimating the survivor function**: When there is no censoring, the survivor function can be estimated as

$$\hat{S}(t) = \frac{n_t}{n}; \ n_t = \text{number of individuals with survival times} \geq t;$$
$$n = \text{the number of individuals in the data set}$$

Kaplan–Meier estimator of the survivor function: Used when the data contain censored observations; based on the ordered survival times, $t_{(1)} \leq t_{(2)} \leq \ldots \leq t_{(n)}$ where $t_{(j)}$ is the jth largest unique survival time. The estimator is given by

$$\hat{S}(t) = \prod_{j|t_{(j)} \leq t} \left(1 - \frac{d_j}{r_j} \right)$$

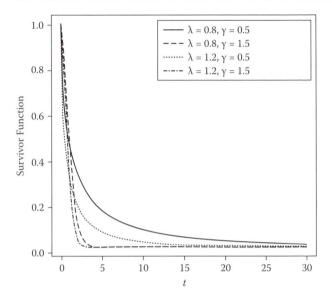

FIGURE 8.2   Survivor functions for a number of Weibull distributions.

where r_j is the number of individuals at risk just before $t_{(j)}$ (including those censored at $t_{(j)}$) and d_j is the number of individuals who experience the terminal event at $t_{(j)}$. For example, the survivor function at the second death time, $t_{(2)}$, is equal to the estimated probability of not dying at $t_{(1)}$ multiplied by the estimated probability, given the individual is still at risk at time $t_{(2)}$, of not dying at time $t_{(2)}$.

Variance of the Kaplan–Meier estimator: Used in constructing confidence intervals for the survivor function and given by

$$\text{Var}\left[\hat{S}(t)\right] = \left[\hat{S}(t)\right]^2 \sum_{j|t_{(j)} \le t} \frac{d_j}{r_j(r_j - d_j)}$$

EXAMPLE 8.1

The data below show the survival times in days of two groups of patients suffering from gastric cancer; members of group 1 received chemotherapy and radiation treatment; members of group 2 received only chemotherapy. Censored observations are indicated by *. Find the survivor function of group 1. In addition, plot the survivor functions of groups 1 and 2 on the same diagram.

GASTRIC CANCER DATA					
GROUP 1			GROUP 2		
17	185	542	1	383	778
42	193	567	63	383	786
44	195	577	105	388	797
48	197	580	125	394	955
60	208	795	182	408	968
72	234	855	216	460	977
74	235	1174*	250	489	1245
95	254	1214	262	523	1271
103	307	1232*	301	524	1420
108	315	1366	301	535	1460*
122	401	1455*	342	562	1516*
144	445	1585*	354	569	1551
167	464	1622*	356	675	1690*
170	484	1626*	358	676	1694
183	528	1736*	380	748	

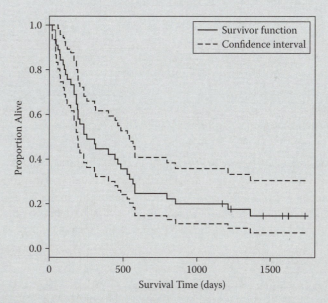

FIGURE 8.3 Kaplan–Meier estimated survivor function for the members of group 1 in the gastric cancer data.

SOLUTION

The plot of the Kaplan–Meier estimate of the survivor function of group 1 is shown in Figure 8.3. Also shown in the plot are the upper and lower 95% confidence interval. (NB: Ticks on the survivor function indicate censored observations.)

The plots of the survivor function of both groups are shown in Figure 8.4; here no confidence limits are given to prevent the diagram looking too 'messy'.

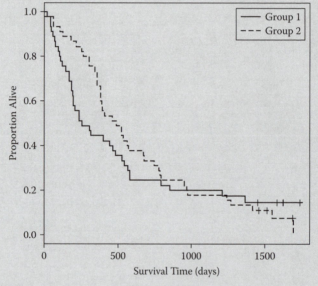

FIGURE 8.4 Kaplan–Meier estimated survivor functions for the members of group 1 and group 2 in the gastric cancer data.

The log-rank test: A test of the equality of two survivor functions from right censored data, i.e., a test of the hypothesis H_0: $S_1 = S_2$. The test is based on a comparison of the observed number of deaths occurring at each death time point (either one or zero) in each group with the number to be expected if the survival experience of the two groups is the same, i.e., if H_0 is true.

EXAMPLE 8.2

The data below give the survival times of five patients in each of two groups. Use the log-rank test to assess the hypothesis, H_0: $S_1 = S_2$. (Right censored observations are indicted by *.)

SURVIVAL TIMES (MONTHS)		
PATIENT	GROUP	SURVIVAL TIME
1	1	2.3
2	1	4.8*
3	1	6.1
4	1	15.2*
5	1	23.8*
6	2	1.6
7	2	3.8
8	2	14.3*
9	2	18.7
10	2	36.3*

SOLUTION

CALCULATING THE LOG-RANK TEST			
TIME			
1.6	Group 1	Group 2	Total
Dead	0(0.5)	1(0.5)	1
Alive	5	4	9
Total	5	5	10
2.3	Group 1	Group 2	Total
Dead	1(0.55)	0 (0.45)	1
Alive	4	4	8
Total	5	4	9
3.8	Group 1	Group 2	Total
Dead	0(0.5)	1(0.5)	1
Alive	4	3	7
Total	4	4	8
6.1	Group 1	Group 2	Total
Dead	1(0.5)	0(0.5)	1
Alive	2	3	5
Total	3	3	6
15.2	Group 1	Group 2	Total
Dead	1(0.5)	0(0.5)	1
Alive	1	2	3
Total	2	2	10

18.7	Group 1	Group 2	Total
Dead	0(0.33)	1(0.67)	1
Alive	1	1	2
Total	1	2	3

Expected number of deaths is shown in parentheses. Alive means alive and at risk.

Observed number of deaths for group 1 = 3 (O_1)
Expected number of deaths for group 1 = 2.89 (E_1)
Observed number of deaths for group 2 = 3 (O_2)
Expected number of deaths for group 2 = 3.11 (E_2)
Test statistic is

$$X^2 = \frac{(O_1 - E_1)^2}{E_1} + \frac{(O_2 - E_2)^2}{E_2}$$

Under H_0 this has a chi-square distribution with a single degree of freedom. Here $X^2 = 0.008$; there is no evidence that the survivor functions of the two groups differ.

The hazard function, $h(t)$: The limiting value of the probability that an individual's survival time, T, lies between t and $t + \delta t$ conditional on T being greater than or equal to t, divided by the time interval δt as δt tends to zero.

$$h(t) = \lim_{\delta t \to 0} \left\{ \frac{\Pr(t \leq T < t + \delta t \mid T \geq t)}{\delta t} \right\}$$

Also known as the *instantaneous failure rate* or *age-specific failure rate*. A measure of how likely an individual is to suffer a terminal event (e.g., death) as a function of the age of the individual.

Estimation of the hazard function: Estimated as the proportion of individuals experiencing the terminal event in an interval per unit time given that they have survived to the beginning of the interval; that is,

$$\hat{h}(t) = \frac{\text{number of individuals experiencing the terminal event in the interval beginning at } t}{(\text{number of individuals surviving at } t) \times (\text{interval width})}$$

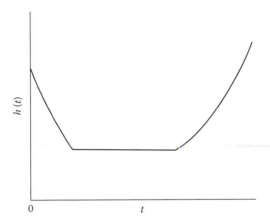

FIGURE 8.5 Bathtub hazard function for death in humans.

The hazard function may increase, decrease, remain constant or be rather more complicated; for example, the hazard function for death in humans has the 'bathtub' shape shown in Figure 8.5.

Relationship between survivor and hazard functions: The hazard function can be defined in terms of the cumulative distribution, $F(t)$, and the probability density function, $f(t)$, of the survival times as follows:

$$h(t) = \frac{f(t)}{1 - F(t)} = \frac{f(t)}{S(t)}$$

It then follows that

$$h(t) = -\frac{d}{dt}\{\log S(t)\}$$

and so

$$S(t) = \exp\{-H(t)\}$$

where $H(t)$ is the integrated or cumulative hazard given by

$$H(t) = \int_0^t h(u)du$$

Hazard functions for the exponential and Weibull distributions: For the exponential

$$h(t) = \frac{\lambda e^{-\lambda t}}{e^{-\lambda t}} = \lambda$$

Here the hazard function is a constant; the hazard of death at any time after the time origin of the study remains the same no matter how much time has elapsed. For the Weibull distribution

$$h(t) = \frac{\lambda \gamma t^{\gamma-1} \exp(-\lambda t^\gamma)}{\exp(-\lambda t^\gamma)} = \lambda \gamma t^{\gamma-1}$$

Weibull hazard functions can be increasing, decreasing or constant depending on the values given to the parameters λ and γ.

8.3 PROPORTIONAL HAZARDS AND COX REGRESSION

Proportional hazards: The hazard functions of two populations, $h_0(t)$ and $h_1(t)$, are proportional if $h_1(t) = \psi h_0(t)$ where ψ is a constant giving the ratio of the hazards of death at any time for an individual in one population relative to an individual in the other population; ψ is known as the *relative hazard* or the *hazard ratio*. Proportional hazards can also be expressed as $\log[h_1(t)] = \log(\psi) + \log[h_0(t)]$.

Proportional hazards imply that if graphs were drawn of $\log[h_1(t)]$ and $\log[h_0(t)]$ then regardless of how complex (or indeed, how simple) say, $h_0(t)$, was, the vertical distance between the two curves at any point in time will be log (ψ). An implication of the proportional hazards assumption is that the population survivor functions for the two groups do not cross.

A simple proportional hazards model: As ψ cannot be negative we can write it as $\exp(\beta)$ where the parameter β is the log of the hazard ratio. By introducing an explanatory variable, x_i, for the ith individual, with values one and zero depending on which population the individual is from, then the hazard function for this individual, $h_i(t)$, can be written as

$$h_i(t) = e^{\beta x_i} h_0(t)$$

Proportional hazards model with multiple covariates: The simple proportional hazards model can be extended to the situation where there are p

covariates measured at the start of the study and which for the ith individual take the values $\mathbf{x}_i' = [x_{i1}, x_{i2}, \ldots, x_{ip}]$ to become the following model:

$$h_i(t) = e^{[\beta_1 x_{i1} + \beta_2 x_{i2} + \ldots + \beta_p x_{ip}]} h_0(t)$$

In this model the regression coefficient, $\exp(\beta_j)$, gives the relative hazard for two individuals differing by one unit on the jth covariate, with all other covariates being the same for the two individuals. In this model, $h_0(t)$ is known as the *baseline hazard function* and is the hazard function for an individual with zero values for all covariates, or if the covariates are re-expressed as differences from their mean values, the hazard function of an individual with the mean value of each covariate. The model can be written in the form

$$\log\left[\frac{h_i(t)}{h_0(t)}\right] = \beta_1 x_{i1} + \beta_2 x_{i2} + \ldots + \beta_p x_{ip}$$

So the proportional hazards model may be regarded as a linear model for the logarithm of the hazard ratio.

Cox's proportional hazards model: A proportional hazards model in which the parameters can be estimated without making any assumptions about the form of the baseline hazard and therefore inferences about the effects of the covariates on the relative hazard can be made without the need for an estimate of $h_0(t)$. Cox's regression is a semi-parametric model; it makes a parametric assumption concerning the effect of the predictors on the hazard function, but makes no assumption regarding the nature of the hazard function itself. In many situations, either the form of the true hazard function is unknown or it is complex and most interest centres on the effects of the covariates rather than the exact nature of the hazard function. Cox's regression allows the shape of the hazard function to be ignored when making inferences about the regression coefficients in the model.

Estimation for Cox's regression: Assume first that there are no tied survival times and that $< t_{(1)} < t_{(2)} < \ldots < t_{(k)}$ represent the k distinct times to the terminal event among n individual times. The conditional probability that an individual with covariate vector x_i suffers a terminal event at time $t_{(i)}$ given that there is a single terminal event at this time and given the risk set R_i (indices of individuals at risk just prior to $t_{(i)}$), is the ratio of the hazards

$$\frac{\exp(\boldsymbol{\beta}'\mathbf{x}_i)}{\sum_{j \in R_i} \exp(\boldsymbol{\beta}'\mathbf{x}_j)}$$

Multiplying these probabilities together for each of the k distinct survival times gives the following partial likelihood function:

$$l(\beta) = \prod_{i=1}^{k} \frac{\exp(\beta' x_i)}{\sum_{j \in R_i} \exp(\beta' x_j)}$$

(NB: The partial likelihood is a function of β only and does not depend on the baseline hazard.)

Maximization of the partial likelihood yields estimates of the parameters with properties similar to those of maximum likelihood estimators.

EXAMPLE 8.3

The data below show the survival times of 20 leukaemia patients, along with the values of three covariates: age at diagnosis, percentage of absolute marrow leukaemia infiltrate and percentage labelling index of the bone marrow leukaemia cells. Also shown is the patient status at the end of the study (0 = dead, 1 = alive). Fit a Cox proportional hazards model to the data and interpret the estimated parameters.

AGE (YEARS)	PERCENTAGE INFILTRATE	PERCENTAGE LABELLING INDEX	SURVIVAL TIME (MONTHS)	STATUS
20	39	7	18	0
25	61	16	31	1
26	55	12	31	0
31	72	5	20	1
36	55	14	8	0
40	91	9	15	0
45	47	14	33	1
50	66	19	12	0
57	83	19	1	1
61	37	4	3	1
63	27	5	1	0
65	75	10	2	1
71	63	11	8	0
71	33	4	3	0
74	58	10	14	1
74	30	16	3	0
75	60	17	13	0

77	69	9	13	0
80	73	7	1	0
82	54	7	6	0

SOLUTION

Fitting the Cox model gives the following results:

	PARAMETER ESTIMATE	STANDARD ERROR (SE)	ESTIMATE/SE
Age	0.042	0.032	1.32
Per. Infil.	0.051	0.036	1.40
Per. Lab.	−0.016	0.122	−0.13

Each additional year of life is estimated to increase the log of the hazard by 0.042; exponentiating the estimated regression coefficient gives the value 1.04, so the hazard function of death for a person aged $x + 1$ years is 1.04 that of a person of age x. The 95% confidence interval for the relative risk is [0.98,1.11]. There is no evidence that age affects survival time of these patients. The other regression coefficients are interpreted in a similar fashion. Testing the hypothesis that all three regression coefficients are zero can be undertaken using three possible tests; the results are

Likelihood ratio test = 3.81 on 3 df, $p = 0.28$
Wald test = 2.72 on 3 df, $p = 0.44$
Score (log-rank) test = 3.44 on 3 df, $p = 0.33$

8.4 DIAGNOSTICS FOR COX REGRESSION

Cox–Snell residual: Defined as $r_i^{(CS)} = \exp(\beta' x_i)\hat{H}_0(t_i)$ where $\hat{H}_0(t_i)$ is the integrated hazard function at time t_i the observed survival time of the ith subject. If the model is correct these residuals have an exponential distribution with mean 1.

Martingale residual: The difference between the event indicator δ_i (equal to 1 if the ith subject died and zero otherwise) and the Cox–Snell residual, $r_i^{(M)} = \delta_i - r_i^{(CS)}$. Martingale residuals take values between $-\infty$ and 1. In a large sample they are uncorrelated and have an expected value of zero. These residuals are not distributed symmetrically around zero.

Deviance residuals: Defined as $r_i^{(D)} = \text{sign}(r_i^{(M)})(\sqrt{-2\{r_i^{(M)} + \delta_i \log(\delta_i - r_i^{(M)})\}})$ Symmetrical about zero when the fitted model is appropriate.

Score residual (Schoenfeld residual): The first derivative of the partial log-likelihood function with respect to one of the explanatory variables. For the jth explanatory variable

$$r_{ij}^{(S)} = \delta_i \left\{ x_{ij} - \frac{\sum_{k \in R(t_i)} x_{kj} \exp(\boldsymbol{\beta}' \mathbf{x}_k)}{\sum_{k \in R(t_i)} \exp(\boldsymbol{\beta}' \mathbf{x}_k)} \right\}$$

where $R(t_i)$ is the set of all individuals at risk at time t_i.

(NB: There is no single value of the score residual for each individual but a set of values, one for each explanatory variable in the fitted model.)

Plots of the re-scaled residuals against time should form a horizontal line if the proportional hazards assumption is valid.

Some residual plots for the Cox model fitted to the leukaemia survival times in Example 8.3 are given in Figures 8.6 and 8.7. The deviance residuals give no concerns about the fitted model but the plots in Figure 8.6 suggest that the proportional hazards assumption may be suspect here, but with such a small data set it is difficult to draw any very convincing conclusion from the plots.

8.5 SUMMARY

Survivor function: $S(t) = \Pr(T > t)$

Kaplan–Meier estimate of survivor function: $\hat{S}(t) = \prod_{j|t_{(j)} \leq t} \left(1 - \frac{d_j}{r_j}\right)$

Variance of the Kaplan–Meier estimator:

$$\text{Var}\left[\hat{S}(t)\right] = \left[\hat{S}(t)\right]^2 \sum_{j|t_{(j)} \leq t} \frac{d_j}{r_j(r_j - d_j)}$$

Hazard function: $h(t) = \lim_{\delta t \to 0} \left\{ \frac{\Pr(t \leq T < t + \delta t \mid T \geq t)}{\delta t} \right\}$

Estimation of hazard function:

$$\hat{h}(t) = \frac{\text{number of individuals experiencing the terminal event in the interval beginning at } t}{(\text{number of individuals surviving at } t) \times (\text{interval width})}$$

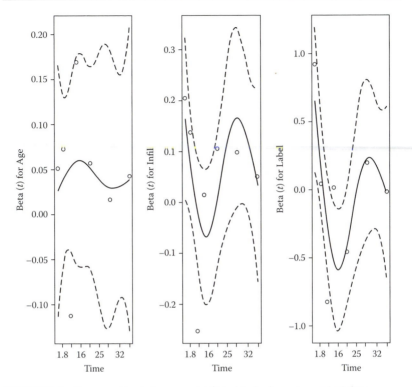

FIGURE 8.6   Score residuals from model fitted to leukaemia survival data.

Integrated or cumulative hazard: $H(t) = \int_0^1 h(u)du$

Proportional hazards: $\log[h_1(t)] = \log(\psi) + \log[h_0(t)]$

Proportional hazards model: $\log\left[\dfrac{h_i(t)}{h_0(t)}\right] = \beta_1 x_{i1} + \beta_2 x_{i2} + \ldots + \beta_p x_{ip}$

Cox–Snell residual: $r_i^{(CS)} = \exp(\boldsymbol{\beta}'\mathbf{x}_i)\,\hat{H}_0(t_i)$

Martingale residual: $r_i^{(M)} = \delta_i - r_i^{(CS)}$

Deviance residual: $r_i^{(D)} = \operatorname{sign}(r_i^{(M)})\left(\sqrt{-2\{r_i^{(M)} + \delta_i \log(\delta_i - r_i^{(M)})\}}\right)$

Score residual: $r_{ij}^{(S)} = \delta_i\left\{x_{ij} - \dfrac{\Sigma_{k \in R(t_i)} x_{kj} \exp(\boldsymbol{\beta}'\mathbf{x}_k)}{\Sigma_{k \in R(t_i)} \exp(\boldsymbol{\beta}'\mathbf{x}_k)}\right\}$

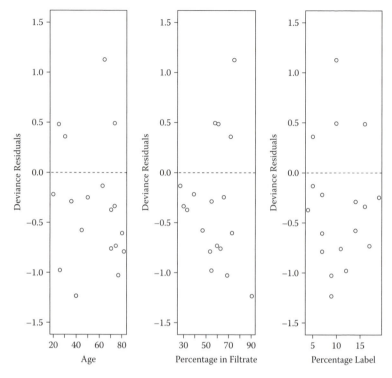

FIGURE 8.7   Deviance residuals from model fitted to leukaemia survival data.

SUGGESTED READING

Collett, D (2004) *Modelling Survival Data in Medical Research,* 2nd edition, Chapman & Hall/CRC, London.

Hosmer, DW and Lemeshow, S (1999) *Applied Survival Analysis*, Wiley, New York.

Longitudinal Data and Their Analysis

9

9.1 LONGITUDINAL DATA AND SOME GRAPHICS

Longitudinal data: Data arising when a response variable of interest is measured/observed on an individual at a number of different time points and a number of explanatory variables are also recorded on each individual. An example is a clinical trial in which, post-randomization, an outcome variable is recorded at weekly intervals for three months. Analysis of such data normally begins with the construction of some informative graphics. (NB: Longitudinal data are also often known as *repeated measures data*.)

Spaghetti plot: A plot on individual subjects' profiles of response values in a longitudinal study, often with an explanatory variable such as treatment group identified.

EXAMPLE 9.1

Two groups of subjects, one with moderate and the other with severe dependence on alcohol, had their salsolinol excretion levels measured (in mmol) on four consecutive days. Construct a spaghetti plot of the data differentiating on the plot between members of the two groups.

SUBJECT		DAY			
		1	2	3	4
Group 1	1	0.33	0.70	2.33	3.20
	2	5.30	0.90	1.80	0.70

	3	2.50	2.10	1.12	1.01
	4	0.98	0.32	3.91	0.56
	5	0.39	0.69	0.73	2.45
	6	0.31	6.34	0.63	3.86
Group 2	1	0.64	0.70	1.00	1.40
	2	0.73	1.85	3.60	2.60
	3	0.70	4.20	7.30	5.40
	4	0.40	1.60	1.40	7.10
	5	2.60	1.30	0.70	0.70
	6	7.80	1.20	2.60	1.80
	7	1.90	1.30	4.40	2.80
	8	0.50	0.40	1.10	8.10

SOLUTION

We can plot all the response profiles on the same diagram using different line types for members of the two groups (see Figure 9.1).

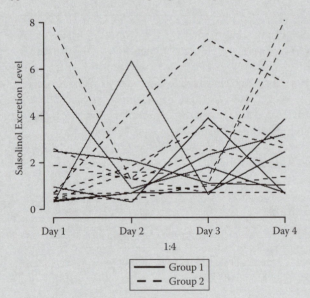

FIGURE 9.1   Spaghetti plot of salsolinol.

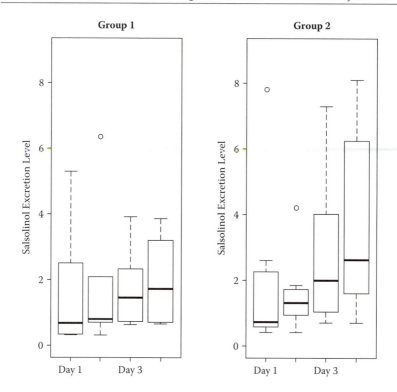

FIGURE 9.2 Boxplots of salsolinol excretion levels for each day for both groups of individuals.

Boxplots of longitudinal data: An alternative to the spaghetti plot, which can become very confused when there are many individual profiles to plot, is boxplots of the response variable values at each time point, again perhaps separating any groups in the data. Such a plot for the salsolinol data in Example 9.1 is shown in Figure 9.2.

Mean profile plot: A plot of the mean profiles and standard errors for groups of individuals in a longitudinal study, for example, treatment and placebo groups in a clinical trial.

EXAMPLE 9.2

The data below come from a clinical trial of a treatment for post-natal depression. Construct the mean profile plot for the two groups.

GROUP	PRE-TREATMENT DEPRESSION	MONTHS POST-TREATMENT					
		1	2	3	4	5	6
1	18	17	18	15	17	14	16
1	25	26	23	18	17	12	10
1	19	17	14	12	13	10	8
1	24	14	23	17	13	12	12
1	19	12	10	8	4	5	5
2	21	13	12	9	9	13	6
2	27	8	17	15	7	5	7
2	24	8	12	10	10	6	5
2	28	14	14	13	12	18	15
2	19	15	16	11	14	12	8

Note: 1 = placebo, 2 = active treatment

SOLUTION

The required plot is shown in Figure 9.3.

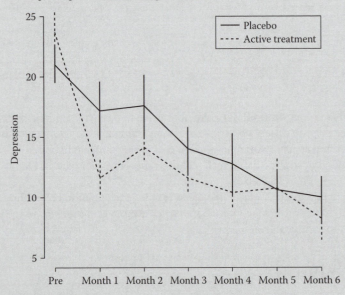

FIGURE 9.3 Mean profiles of placebo and treatment groups showing standard errors at each time point.

Scatterplot matrix: An arrangement of the pairwise scatter plots of the repeated measurements in a set of longitudinal data. Each panel of the matrix is a scatter plot of one variable against another. Useful in assessing how the measurements at various time points are related, information needed when it comes to modelling the longitudinal data. The scatterplot matrix of the salsolinol data is shown in Figure 9.4.

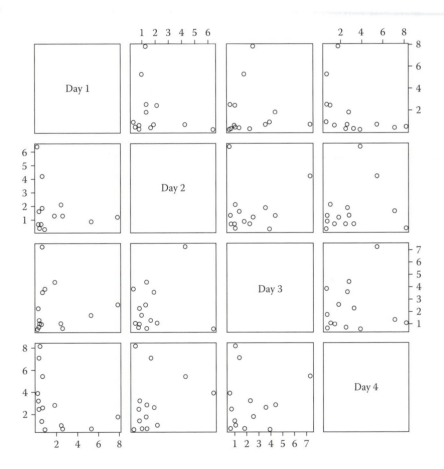

FIGURE 9.4   Scatterplot matrix of salsolinol excretion levels.

9.2 SUMMARY MEASURE ANALYSIS

Summary measure analysis (response feature analysis): A simple approach to the analysis of longitudinal data that involves the calculation of a suitable summary measure (or measures) from the set of repeated response values available on each individual. A commonly used summary measure is the mean of the repeated measurements.

EXAMPLE 9.3

Compare the salsolinol excretion levels of the two groups of alcoholics for the data in Example 9.1 using the mean of the excretion levels over the four days as the summary measure.

SOLUTION

The means of the day 1 to day 4 measurements of the six individuals in group 1 and the eight individuals in group 2 form the data used to compare the two groups. First they can be compared informally by boxplots of the summary measures of the two groups (see Figure 9.5).

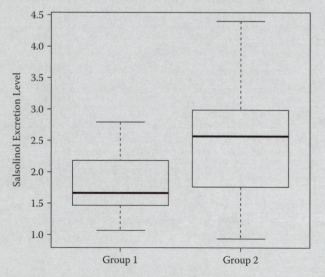

FIGURE 9.5 Boxplots of mean summary measures for the two groups in the salsolinol data.

An independent samples t-test for the mean summary measures gives the following results:

$t = -1.40$, df = 12, p-value = 0.19
95% confidence interval [-1.77, 0.39]
sample estimates: mean of group 1 = 1.80, mean of group 2 = 2.49

There is no evidence that the average salsolinol excretion level over the four days is different in the two groups, although with so few individuals in each group the power of the test is low.

Problems of the summary measure approach: Although simple to apply, the summary measure approach has a number of distinct drawbacks; one such is that it forces the investigator to focus on only a single aspect of the repeated measurements over time. It seems intuitively clear that when T repeated measures are replaced by a single number summary, there must necessarily be some loss of information. And it is possible for individuals with quite different response profiles to have the same or similar values for the chosen summary measure. Finally, the simplicity of the summary measure method is lost when there are missing data or the repeated measures are irregularly spaced, as is the method's efficiency and the efficiency of the method is reduced.

9.3 LINEAR MIXED EFFECTS MODELS

An independence model for longitudinal data: Assuming the independence of the repeated measurements of the response variable and the conditional normality of this variable in a set of longitudinal data implies that the data can be modelled using the multiple linear regression model described in Chapter 6.

EXAMPLE 9.4

For the data in Example 9.2 assume that the monthly measures of depression are independent of one another and fit a multiple regression model to assess the effect of treatment, time and the pre-randomization measure of depression on the response.

SOLUTION

In Example 9.2 the data are given in what is known as the 'wide form'. To make the data suitable for fitting the multiple linear regression model we first need to put them into what is known as the 'long form'; this form of the data is shown below for the first three individuals.

	TREAT	TIME	PRE	DEP
1	Placebo	1	18	17
2	Placebo	2	18	18
3	Placebo	3	18	15
4	Placebo	4	18	17
5	Placebo	5	18	14
6	Placebo	6	18	15
7	Placebo	1	25	26
8	Placebo	2	25	23
9	Placebo	3	25	18
10	Placebo	4	25	17
11	Placebo	5	25	12
12	Placebo	6	25	10
13	Placebo	1	19	17
14	Placebo	2	19	14
15	Placebo	3	19	12
16	Placebo	4	19	13
17	Placebo	5	19	10
18	Placebo	6	19	8

In all, the long form of the data has 60 rows, each one corresponding to the values of the covariates and the response at one of the time periods. Assuming independence of the repeated measure means that we simply analyse the data as though it consisted of 60 separate observations. Fitting a multiple regression model to the data with depression score as the response variable and time, treatment (coded 0 and 1 for placebo and active) and pre-randomization depression as explanatory variables gives the following results:

	ESTIMATE	STANDARD ERROR (SE)	ESTIMATE/SE
Intercept	11.27	3.55	3.18
Treatment	-3.47	1.10	-3.16
Time	-1.24	0.29	-4.21
Pre	0.32	0.16	2.04

> Multiple $R^2 = 0.34$
> F-statistic for testing that all regression coefficients are zero takes the value 9.45 and has 3 and 56 df.

(NB: Independence of the repeated measurements of the response is, for most longitudinal data sets, totally unrealistic. Because several observations of the response variable are made on the same individual, it is likely that the measurements will be correlated rather than independent, even after conditioning on the explanatory variables.)

Random effects models: Suitable models for longitudinal data need to include parameters analogous to the regression coefficients in the usual multiple regression model that relate the explanatory variables to the repeated measurements, *and*, in addition, parameters that account adequately for the correlational structure of the repeated measurements of the response variable. It is the regression coefficients that are generally of most interest, with the correlational structure parameters often being regarded as *nuisance parameters*; but providing an adequate model for the correlational structure of the repeated measures *is* necessary to avoid misleading inferences about those parameters that are of most importance to the researcher. Although the estimation of the correlational structure of the repeated measurements is usually regarded as a secondary aspect of any analysis (relative to the mean response over time), the estimated correlational structure must describe the *actual* correlational structure present in the data relatively accurately to avoid making misleading inferences on the substantive parameters. Assuming that the correlation amongst repeated measurements of the response on the same individual arises from shared unobservable variables, random effects models introduce these unobserved variables into the model as random variables, i.e., random effects. So models for longitudinal data will include both fixed and random effects and are generally known as *linear mixed effects models*.

Random intercept model: This is the simplest linear mixed effects model and for a set of longitudinal data in which y_{ij} represents the value of the response for individual i at time t_j

$$y_{ij} = \beta_0 + \beta_1 t_j + u_i + \varepsilon_{ij}$$

i.e.,

$$y_{ij} = (\beta_0 + u_i) + \beta_1 t_j + \varepsilon_{ij}$$

(We are assuming no other covariates for simplicity.) Here the total residual that would be present in the usual linear regression model has been

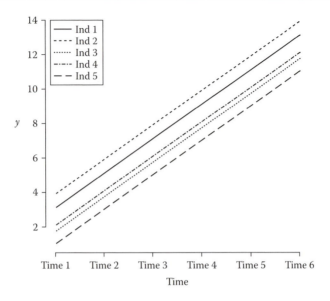

FIGURE 9.6 The random intercept model.

portioned into a subject-specific random component, u_i, which is constant over time, plus a residual, ε_{ij}, that varies randomly over time. We assume $u_i \sim N(0, \sigma_1^2)$ and $\varepsilon_{ij} \sim N(0, \sigma^2)$ and that the two terms are independent of each other and of the time t_j. The model allows the regression lines of the individuals to have random intercepts, whereas time has a fixed effect, β_1. The model is illustrated in Figure 9.6 which shows the fitted regression lines for five individuals observed at six time points.

In the random intercept model

$$\text{Var}(y_{ij}) = \text{Var}(u_i + \varepsilon_{ij}) = \sigma_1^2 + \sigma^2$$

$$\text{Cov}(u_i + \varepsilon_{ij}, u_i + \varepsilon_{ik}) = \sigma_1^2$$

The model constrains the variance of each repeated measurement to be the same and the covariance between any pair of measurements to be equal. This is usually known as the *compound symmetry* structure. This structure is often not realistic for longitudinal data, where it is more common for measures taken closer together in time to be more highly correlated than those taken further apart.

Random intercept and slope model: A linear mixed effects model which allows both random intercepts and random slopes, which allows a more realistic covariance structure for the repeated measurements. The model here is

$$y_{ij} = (\beta_0 + u_i) + (\beta_1 + v_i)t_j + \varepsilon_{ij}$$

The two random effects u and v are assumed to have a bivariate normal distribution with zero means for both variables, variances σ_1^2 and σ_2^2 and covariance σ_{12}. This model allows both random intercepts and random slopes. The random intercept and slope model is illustrated in Figure 9.7.

In the random intercept and slope model

$$\mathrm{Var}(y_{ij}) = \sigma_1^2 + 2\sigma_{12}t_j + \sigma_2^2 t_j^2 + \sigma^2$$

$$\mathrm{Cov}(u_i + v_i t_j + \varepsilon_{ij}, u_i + v_i t_k + \varepsilon_{ik}) = \sigma_1^2 + \sigma_{12}(t_j - t_k) + \sigma_2^2 t_j t_k$$

With this model the variances of the response at each time point are not constrained to be the same and neither are the pairwise covariances.

Models can be compared by means of a likelihood ratio test.

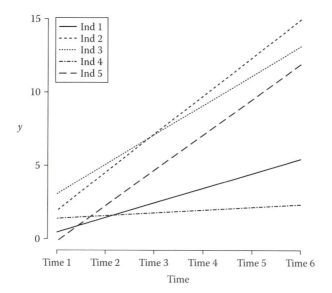

FIGURE 9.7   Random intercept and slope model.

Estimation of the parameters in linear mixed effects models: If the general linear mixed effects model is written as

$$E(y_i) = X_i\beta$$

where y_i is the vector of repeated response values and X_i the design matrix for the ith individual with the covariance matrix of the y_i being

$$Cov(y_i) = \Sigma_i(\theta)$$

where θ is a vector of parameters of covariance parameters, then the generalized least squares estimator of β is

$$\hat{\beta} = \frac{\sum_{i=1}^{n}\left(X_i'\Sigma_i(\hat{\theta})y_i\right)}{\sum_{i=1}^{n}\left(X_i'\Sigma_i(\hat{\theta})X_i\right)}$$

where $\hat{\theta}$ is the maximum likelihood estimator of θ.

EXAMPLE 9.5

Fit both the random intercept and random intercept and slope models to the data in Example 9.4

SOLUTION

The results from fitting both models is as follows:
    Random intercept
    Random effects:

$$\hat{\sigma}_1^2 = 9.69, SD = 3.11$$

$$\hat{\sigma}^2 = 7.86, SD = 2.80$$

FIXED EFFECTS		
		STANDARD	
	ESTIMATE	ERROR (SE)	ESTIMATE/SE
Intercept	11.27	7.13	1.58
Treatment	−3.47	2.29	−1.51
Time	−1.24	0.21	−5.84
Pre	0.32	0.33	0.98

Random intercept and slope model
Random effects:

FIXED EFFECTS		
		STANDARD	
	ESTIMATE	ERROR (SE)	ESTIMATE/SE
Intercept	11.21	7.14	1.57
Treatment	−3.69	2.29	−1.61
Time	−1.24	0.30	−4.19
Pre	0.33	0.33	1.00

The log-likelihood comparing the model gives a chi-square test statistic value of 2.93 on 2 df; the random intercept and slope model does not produce a significant increase in fit over the random intercept model. Note that for the treatment effect the standard error in the random intercept model is about twice that in the independence model fitted in Example 9.4, with the result that the treatment effect is no longer significantly different from zero.

9.4 MISSING DATA IN LONGITUDINAL STUDIES

Missing data: Planned observations that were not made for some reason. In longitudinal data missing values may occur because individuals drop out of the study completely or do not appear for one or other scheduled visits. Missing values in longitudinal data complicate the analysis.

The missing-data mechanism: Concerns the reasons why values are missing, in particular whether these reasons relate to recorded (non-missing) values for a subject. Missing-data mechanisms are usually classified into three types and the type of mechanism involved has implications for which approaches to analysis are suitable and which are not. The classification is as follows:

- *Missing completely at random* (MCAR): The missing-data mechanism when missingness does not depend on the values of the data values in **Y**, missing or observed, i.e., $f(\mathbf{M}|\mathbf{Y}, \varphi) = f(\mathbf{M}|\varphi)$ where $f(\mathbf{M}|\mathbf{Y}, \varphi)$ is the conditional distribution of **M** given **Y** and φ is a vector of unknown parameters. Note that this assumption does not mean that the pattern itself is random but only that missingness does not depend on the data values either observed or unobserved. Consequently the observed (non-missing) values effectively constitute a simple random sample of the values for all subjects. The classification of missing values as MCAR implies that Pr(missing|observed,unobserved) = Pr(missing). Possible examples include missing laboratory measurements because of a dropped test tube (if it was not dropped because of the knowledge of any measurement), the accidental death of a participant in a study, or a participant moving to another area. Intermittent missing values in a longitudinal data set, whereby a patient misses a clinic visit for transitory reasons ("went shopping instead" or the like) can reasonably be assumed to be MCAR. When data are MCAR, missing values are no different than non-missing in terms of the analysis to be performed, and the only real penalty in failing to account for the missing data is loss of power. But MCAR is a strong assumption because missing *does* usually depend on at least the observed/recorded values.
- *Missing at random* (MAR): The missing at random missing-value mechanism occurs when the missingness depends only on observed values, $\mathbf{Y}_{obs}$, and not on values that are missing, $\mathbf{Y}_{miss}$, i.e., $f(\mathbf{M}|\mathbf{Y}, \varphi) = f(\mathbf{M}|\mathbf{Y}_{obs}, \varphi)$, for all $\mathbf{Y}_{miss}, \varphi$. Here missingness depends only on the observed data, with the distribution of future values for a subject who drops out at a particular time being the same as the distribution of the future values of a subject who remains in at that time, if they have the same covariates and the same past history of outcome up to and including the specific time point. In classifying missing values as MAR we imply that Pr(missing|observed,unobserved) = Pr(missing|observed). This type of missing value is also called *ignorable* because conclusions based on likelihood methods are not affected by MAR data.

  An example of a MAR mechanism is provided by a study in which the response measure is body mass index (BMI). Suppose that the measure is missing because subjects who had high body mass index values at earlier visits avoided being measured at later visits out of embarrassment, regardless of whether they had gained or lost weight in the intervening period. The missing values here are MAR but not MCAR; consequently, methods applied to the data that assumed the latter might give misleading results (see later discussion). In this case, missing data depend on known values and thus are described fully by variables observed in the data set. Accounting for the values which "cause" the missing data will produce unbiased results in an analysis.

- *Non-ignorable* (sometimes referred to as *informative*): The final type of dropout mechanism is one where the missingness depends on the unrecorded missing values–observations are likely to be missing when the outcome values that would have been observed had the patient not dropped out are systematically higher or lower than usual (corresponding perhaps to their condition becoming worse or improving). An example is a participant dropping out of a longitudinal study when his/her blood pressure became very high and this value is not observed, or when the pain becomes intolerable and the associated pain value is not recorded. In the BMI example introduced above, if subjects were more likely to avoid being measured if they had put on extra weight since the last visit, then the data are non-ignorably missing. Dealing with data containing missing values that result from this type of missing-data mechanism is difficult. The correct analyses for such data must estimate the dependence of the missingness probability on the missing values.

9.5 SUMMARY

Longitudinal data: Data arising when a response variable of interest is measured/observed on an individual at a number of different time points.

Spaghetti plot: A plot on individual subjects' profiles of response values in a longitudinal study, often with an explanatory variable such as treatment group identified.

Summary measure analysis (response feature analysis): A simple approach to the analysis of longitudinal data that involves the calculation of a suitable summary measure (or measures) from the set of repeated response values available on each individual.

Random intercept model: $y_{ij} = (\beta_0 + u_i) + \beta_1 t_j + \varepsilon_{ij}$

$$\mathrm{Var}(y_{ij}) = \mathrm{Var}(u_i + \varepsilon_{ij}) = \sigma_1^2 + \sigma^2$$

$$\mathrm{Cov}(u_i + \varepsilon_{ij}, u_i + \varepsilon_{ik}) = \sigma_1^2$$

Random intercept and slope model: $y_{ij} = (\beta_0 + u_i) + (\beta_1 + v_i)t_j + \varepsilon_{ij}$

$$\mathrm{Var}(y_{ij}) = \sigma_1^2 + 2\sigma_{12}t_j + \sigma_2^2 t_j^2 + \sigma^2$$

$$\mathrm{Cov}(u_i + v_i t_j + \varepsilon_{ij}, u_i + v_i t_k + \varepsilon_{ik}) = \sigma_1^2 + \sigma_{12}(t_j - t_k) + \sigma_2^2 t_j t_k$$

Estimation of fixed effects: $\displaystyle \hat{\beta} = \frac{\sum_{i=1}^{n}\left(\mathbf{X}_i' \mathbf{\Sigma}_i(\hat{\theta})\mathbf{y}_i\right)}{\sum_{i=1}^{n}\left(\mathbf{X}_i' \mathbf{\Sigma}_i(\hat{\theta})\mathbf{X}_i\right)}$

Missing completely at random (MCAR): The missing-data mechanism when missingness does not depend on the values of the data values in $\mathbf{Y}$, missing or observed, i.e., $f(\mathbf{M}|\mathbf{Y}, \varphi) = f(\mathbf{M}|\varphi)$, where $f(\mathbf{M}|\mathbf{Y}, \varphi)$ is the conditional distribution of $\mathbf{M}$ given $\mathbf{Y}$ and φ is a vector of unknown parameters.

Missing at random (MAR): The missing at random missing-value mechanism occurs when the missingness depends only on observed values, $\mathbf{Y}_{obs}$, and not on values that are missing, $\mathbf{Y}_{miss}$, i.e., $f(\mathbf{M}|\mathbf{Y}, \varphi) = f(\mathbf{M}|\mathbf{Y}_{obs}, \varphi)$ for all $\mathbf{Y}_{obs}, \varphi$.

Non-ignorable (informative): Missingness depends on the unrecorded missing values.

SUGGESTED READING

Fitzmaurice, GM, Laird, NM and Ware, JH (2004) *Applied Longitudinal Analysis*, Wiley, New York.

Rabe-Hesketh, S and Skrondal, A (2005) *Multilevel and Longitudinal Modeling Using Stata*, Stata Press, College Station, Texas.

Twisk, JWR (2003) *Applied Longitudinal Data Analysis for Epidemiology*, Cambridge University Press, Cambridge.

Multivariate Data and Their Analysis

10

10.1 MULTIVARIATE DATA

Multivariate data: Data resulting from the measurement of the values of several random variables on each individual in a sample, leading to a *vector-valued* or *multidimensional* observation for each. An example of a multivariate data set is shown in Table 10.1

The majority of data sets collected by researchers in all disciplines are multivariate. Although in some cases where multivariate data have been collected it may make sense to isolate each variable and study it separately, in the main it does not. Because the whole set of variables is measured on each sample member, the variables will be related to a greater or lesser degree. Consequently, if each variable is analyzed in isolation, the full structure of the data may not be revealed. With the great majority of multivariate data sets all the variables need to be examined simultaneously in order to uncover the patterns and key features in the data and hence the need for techniques which allow for a multivariate analysis of the data.

The multivariate data matrix: An $n \times p$ matrix $\mathbf{X}$. used to represent a set of multivariate data for n individuals each having p variable values:

$$\mathbf{X} = \begin{bmatrix} x_{11} & \cdots & x_{1p} \\ \vdots & \ddots & \vdots \\ x_{n1} & \cdots & x_{np} \end{bmatrix}$$

where x_{ij} represents the variable value on the jth variable for the ith individual.

TABLE 10.1   Chest, waist and hip measurements (in inches) for five individuals

INDIVIDUAL	CHEST	WAIST	HIPS
1	34	30	32
2	37	32	37
3	38	30	36
4	36	33	39
5	38	29	33

10.2 MEAN VECTORS, VARIANCES, COVARIANCE AND CORRELATION MATRICES

Mean vectors: For p variables, the population mean vector is $\mathbf{\mu}' = [\mu_1, \mu_2, \ldots \mu_p]$, where $\mu_i = E(x_i)$ An *estimate of* $\mathbf{\mu}'$, based on n, p-dimensional observations, is $\mathbf{\bar{x}}' = [\bar{x}_1, \bar{x}_2, \ldots, \bar{x}_p]$ where $\bar{x}_i$ is the sample mean of variable x_i.

Variances: The population vector of variances is $\mathbf{\sigma}' = [\sigma_1^2, \sigma_2^2, \ldots, \sigma_p^2]$ where $\sigma_i^2 = E(x_i - \mu_i)^2$ is the variance of the ith variable. An estimate of $\mathbf{\sigma}'$ based on n, p-dimensional observations is $\mathbf{s}' = [s_1^2, s_2^2, \ldots, s_p^2]$ where s_i^2 is the sample variance of x_i.

Covariance matrix: The population covariance of two variables, x_i and x_j, is defined as $\mathrm{Cov}(x_i\, x_j) = E(x_i - \mu_i)\,(x_j - \mu_j)$. If $i = j$, we note that the covariance of the variable with itself is simply its variance, and therefore there is no need to define variances and covariances independently in the multivariate case.

The covariance of x_i and x_j is usually denoted by σ_{ij} (so the variance of the variable x_i is often denoted by σ_{ij} rather than σ_i^2). With p variables, $x_1, x_2, \ldots, x_p$, there are p variances and $p(p-1)/2$ covariances. In general these quantities are arranged in a $p \times p$ symmetric matrix, $\mathbf{\Sigma}$, where

$$\mathbf{\Sigma} = \begin{pmatrix} \sigma_{11} & \sigma_{12} & \cdots & \sigma_{1p} \\ \sigma_{21} & \sigma_{22} & \cdots & \sigma_{2p} \\ \vdots & \vdots & \vdots & \vdots \\ \sigma_{p1} & \sigma_{p2} & \cdots & \sigma_{pp} \end{pmatrix}$$

(NB: $\sigma_{ij} = \sigma_{ji}$.) The matrix Σ is generally known as the *variance-covariance matrix* or simply the *covariance matrix*; it is estimated by the matrix S, given by

$$S = \sum_{i=1}^{n} (\mathbf{x}_i - \bar{\mathbf{x}})(\mathbf{x}_i - \bar{\mathbf{x}})' / (n - 1)$$

where $\mathbf{x}_i' = [x_{i1}, x_{i2}, \ldots x_{ip}]$ is the vector of observations for the ith individual. (NB: The main diagonal of S contains the variances of each variable.)

Correlation matrix: The covariance between two variables is often difficult to interpret because it depends on the units in which the two variables are measured; consequently, it is often standardized by dividing by the product of the standard deviations of the two variables to give the correlation coefficient; the population correlation for two variables ρ_{ij} is

$$\rho_{ij} = \frac{\sigma_{ij}}{\sqrt{\sigma_{ii}\sigma_{jj}}}$$

The correlation coefficient lies between -1 and $+1$ and gives a measure of the *linear* relationship of the variables x_i and x_j. With p variables there are $p(p-1)/2$ distinct correlations which may be arranged in a $p \times p$ matrix whose diagonal elements are unity. The population correlation matrix is generally estimated by a matrix containing the Pearson correlation coefficients for each pair of variables; this matrix is generally denoted by R which can be written in terms of the sample covariance matrix S, as

$$R = D^{-1/2}SD^{-1/2}$$

where $D^{-1/2} = \operatorname{diag}(1/s_i)$ and s_i is the sample standard deviation of variable x_i.

(NB: In general covariance and correlation matrices are of full rank, p, so that both matrices will be non-singular, i.e., invertible, to give matrices S^{-1} or R^{-1}.)

EXAMPLE 10.1

Find the mean vector, the vector of variances, the covariance matrix and the correlation matrix for the data of chest, waist and hip measurements in Table 10.1.

SOLUTION

	CHEST	WAIST	HIPS
Means	36.6	30.8	35.4
Variances	2.8	2.7	8.3
COVARIANCE MATRIX			
Chest	2.8	–0.6	1.2
Waist	–0.6	2.7	4.1
Hips	1.2	4.1	8.3
CORRELATION MATRIX			
Chest	1.00	–0.22	0.25
Waist	–0.22	1.00	0.87
Hips	0.25	0.87	1.00

10.3 TWO MULTIVARIATE DISTRIBUTIONS: THE MULTINOMIAL DISTRIBUTION AND THE MULTIVARIATE NORMAL DISTRIBUTION

Multinomial distribution: The probability distribution of the number of occurrences of mutually exclusive and collectively exhaustive events $E_1, E_2, \ldots, E_k$ with, respectively, probabilities, $p_1, p_2, \ldots, p_k$ in a sequence of independent and identical trials, each of which can result in one of the events and given by

$$\Pr(n_1, n_2, \ldots, n_k) = \frac{n!}{n_1! n_2! \ldots n_k!} p_1^{n_1} p_2^{n_2} \ldots p_k^{n_k}, n_i = 0, 1, \ldots, n; \ n = \sum_{i=1}^{k} n_i$$

Such a multinomial distribution is often designated as $M(n, p_1, p_2, \ldots p_k)$.

$$E(n_i) = np_i; \ \mathrm{Var}(n_i) = np_i(1 - p_i); \ \mathrm{Cov}(n_i, n_j) = -np_i p_j$$

Marginal distribution: The marginal distribution of n_i is the binomial distribution $\dfrac{n!}{n_1!(n-n_1)!}p_1^{n_1}(1-p_1)^{n-n_1}$, $n_i = 0,1,2\ldots,n$

Conditional distribution: The conditional distribution of $(n_{j+1}, n_{j+2}, \ldots, n_k)$ given $(n_1 = m_1, n_2 = m_2, \ldots n_j = m_j)$ is $M(n-m, q_{j+1}, q_{j+2}, \ldots, q_k)$ where $m = m_1 + m_2 + \ldots + m_k$, $q_i = p_i / P$ (for $i = j+1, \ldots, k$) and $P = p_{j+1} + \ldots + p_k$.

Multivariate normal distribution: A distribution for a p-dimensional *random vector*, $\mathbf{X}' = [X_1, X_2, \ldots, X_p]$ given by

$$f(\mathbf{X}) = \frac{1}{(2\pi)^{p/2}} |\Sigma|^{-1/2} \exp\left\{-\frac{1}{2}(\mathbf{X}-\mu)\Sigma^{-1}(\mathbf{X}-\mu)'\right\}; -\infty < X_i < \infty, i = 1 \ldots p$$

where $\mu = E(\mathbf{X}) = \left[E(X_1), E(X_2), \ldots, E(X_p)\right]$ and Σ is the covariance matrix of $\mathbf{X}$.

Such a distribution is often designated as MVN(μ, Σ).

Linear combination of variables: The function L given by

$$L = \sum_{i=1}^{p} c_i X_i = \mathbf{c}'\mathbf{X} \text{ where } \mathbf{c}' = [c_1, c_2, \ldots, c_p] \text{ is a vector of scalars.}$$

If $\mathbf{X} \sim$ MVN(μ, Σ) then $L \sim$ N(μ, σ^2) where $\mu = \mathbf{c}'\mu$ and $\sigma^2 = \mathbf{c}'\Sigma\mathbf{c}$.

Linear transformations: If $\mathbf{C}$ is a $p \times p$ non-singular matrix and $\mathbf{Y} = \mathbf{CX}$ then $\mathbf{Y} \sim$ MVN($\mathbf{C}\mu$, $\mathbf{C}\Sigma\mathbf{C}'$).

Bivariate normal distribution: The multivariate normal distribution when $p = 2$; denoting the two variables by x and y the distribution is given explicitly by

$$f(x,y) = \frac{1}{2\pi\sigma_1\sigma_2\sqrt{(1-\rho^2)}} \exp\left[\frac{(x-\mu_1)^2}{\sigma_1^2} - 2\rho\frac{(x-\mu_1)(y-\mu_2)}{\sigma_1\sigma_2} + \frac{(x-\mu_2)^2}{\sigma_2^2}\right],$$

$$-\infty < x, y < \infty$$

where ρ is the correlation coefficient of the two variables. Some bivariate distributions all with means zero, variances one and different values of the correlation are shown in Figure 10.1.

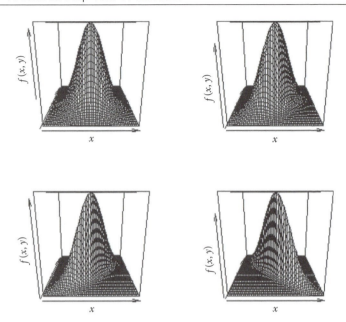

FIGURE 10.1   Four bivariate normal distributions with correlations, 0.0, 0.5, 0.8, −0.9.

10.4 THE WISHART DISTRIBUTION

Wishart distribution: The joint distribution of the variances and covariances in samples of size n from a multivariate normal distribution. Given by

$$f(\mathbf{S}) = \frac{\left(\dfrac{1}{2}n\right)^{\frac{1}{2}p(n-1)} \mid \boldsymbol{\Sigma}^{-1} \mid^{\frac{1}{2}(n-1)} \mid \mathbf{S} \mid^{\frac{1}{2}(n-p-2)}}{\pi^{\frac{1}{4}p(p-1)} \displaystyle\prod_{j=1}^{p} \Gamma\left\{\dfrac{1}{2}(n-j)\right\}} \exp\left\{-\frac{1}{2}n\,\mathrm{trace}(\boldsymbol{\Sigma}^{-1}\mathbf{S})\right\}$$

where $\mathbf{S}$ is the sample covariance matrix with n rather than $n-1$ in the denominator.

10.5 PRINCIPAL COMPONENTS ANALYSIS

Curse of dimensionality: A term applied to the possible problem of having too many variables in a set of multivariate data and thus making the application of a number of statistical techniques to the data more difficult.

Principal components analysis: A multivariate technique that aims to reduce the dimensionality of a multivariate data set while retaining as much of the information in the data as possible. Essentially a transformation of the data to a new set of uncorrelated variables, each of which is a linear function of the original variables and which accounts for decreasing proportions of the variation associated with the original variables.

Technical details of principal components analysis: If $\mathbf{x}$ is a p-dimensional random vector with mean μ and covariance matrix $\mathbf{\Sigma}$, then the principal component transformation to the vector y is the following:

$$\mathbf{x} \rightarrow y = \mathbf{\Gamma}'(\mathbf{x} - \mu)$$

where $\mathbf{\Gamma}$ is orthogonal with columns which are the eigenvectors of $\mathbf{\Sigma}$; $\mathbf{\Gamma}'\mathbf{\Sigma}\,\mathbf{\Gamma} = \mathbf{\Lambda}$ where $\mathbf{\Lambda}$ is a diagonal matrix whose elements are the eigenvalues of $\mathbf{\Sigma}$, $\lambda_1, \lambda_2, \ldots, \lambda_p$ with $\lambda_1 \geq \lambda_2 \geq \ldots \geq \lambda_p$. The ith principal component may be defined as the ith element of the vector y which is given explicitly as

$$y_i = \gamma_i' (\mathbf{x} - \mu)$$

where γ_i' is the ith column of $\mathbf{\Gamma}$ and gives the coefficients defining the ith principal component.

The principal component variables have the following properties:

$$E(y_i) = 0, \operatorname{Var}(y_i) = \lambda_i, \operatorname{Cov}(y_i, y_j) = 0, i \neq j, \operatorname{Var}(y_1) \geq \operatorname{Var}(y_2) \ldots \geq \operatorname{Var}(y_p) \geq 0,$$

$$\sum_{i=1}^{p} \operatorname{Var}(y_i) = \operatorname{trace}(\mathbf{\Sigma}), \prod_{i=1}^{p} \operatorname{Var}(y_i) = |\mathbf{\Sigma}|.$$

The sum of the first k eigenvalues divided by the sum of all eigenvalues, $\dfrac{\lambda_1 + \lambda_2 + \ldots + \lambda_k}{\lambda_1 + \lambda_2 + \ldots + \lambda_p}$ represents the proportion of the total variation explained by the first k principal components.

EXAMPLE 10.2
What are the principal components of bivariate data with covariance matrix

$$\Sigma = \begin{pmatrix} 1 & \rho \\ \rho & 1 \end{pmatrix} \text{ and zero means for both variables?}$$

SOLUTION
The eigenvalues of the covariance matrix are $1 \pm \rho$. If ρ is positive the first eigenvector is $(1,1)$ and the first principal component is

$$y_1 = \frac{1}{\sqrt{2}}(x_1 + x_2)$$

with variance $1 + \rho$. The second principal component is

$$y_2 = \frac{1}{\sqrt{2}}(x_1 - x_2)$$

with variance $1 - \rho$. If ρ is negative then the order of the principal components is reversed. (NB: The sum of the eigenvalues is equal to the sum of the variances of x_1 and x_2.)

Sample principal components: The principal components of a sample of multivariate data are found from the eigenvectors of the sample covariance matrix, although when the variables are on different scales (as they usually will be) it is more sensible to extract the components from the covariance of the standardized variables, i.e., from the correlation matrix.

EXAMPLE 10.3

Find the principal components of the following set of multivariate data which gives crime rates for seven types of crime in 15 states of the United States.

STATE ID	MURDER	RAPE	ROBBERY	ASSAULT	BURGLARY	THEFT	VEHICLE
ME	2.0	14.8	28	102	803	2347	164
NH	2.2	21.5	24	92	755	2208	228
VT	2.0	21.8	22	103	949	2697	181
MA	3.6	29.7	193	331	1071	2189	906

RI	3.5	21.4	119	192	1294	2568	705
CT	4.6	23.8	192	205	1198	2758	447
NY	10.7	30.5	514	431	1221	2924	637
NJ	5.2	33.2	269	265	1071	2822	776
PA	5.5	25.1	152	176	735	1654	354
OH	5.5	38.6	142	235	988	2574	376
IN	6.0	25.9	90	186	887	2333	328
IL	8.9	32.4	325	434	1180	2938	628
MI	11.3	67.4	301	424	1509	3378	800
WI	3.1	20.1	73	162	783	2802	254
MN	2.5	31.8	102	148	1004	2785	288

SOLUTION

Here the seven variables are on very different scales so it makes more sense to extract the components from the correlation matrix rather than the covariance matrix. Details of the principal components are given below.

Standard deviations of components and proportion of variance accounted for

	COMP. 1	COMP. 2	COMP. 3	COMP. 4	COMP. 5	COMP. 6	COMP. 7
Standard deviation	2.2	0.91	0.744	0.669	0.367	0.2484	0.1787
Proportion of variance	0.7	0.12	0.079	0.064	0.019	0.0088	0.0046
Cumulative proportion	0.7	0.82	0.903	0.967	0.987	0.9954	1.0000

The first two components account for over 80% of the variance in the original seven variables. (NB: All seven principal components are needed to account for *all* the variation of the original variables.) A plot of component variances (sometimes know as a *scree plot*) is often helpful in deciding on the number of components to adequately describe the data—a distinct 'elbow' in the curve can be used to indicate the number of components needed. The scree plot for this example is shown in Figure 10.2. Here the plot suggests that two components only are required.

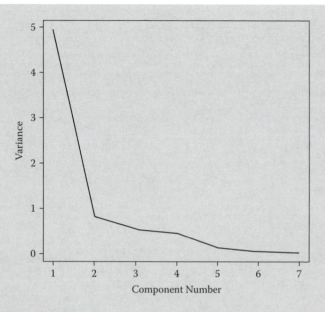

FIGURE 10.2   Scree plot of component variances for city crime data.

The coefficients defining the principal components are given below. The first component is seen to be a weighted average of the standardized crime rates and the second is a contrast of the crime rates for murder, robbery, assault and vehicle against rape, burglary and theft.

	COMP. 1	COMP. 2	COMP. 3	COMP. 4	COMP. 5	COMP. 6	COMP. 7
Murder	−0.400	−0.162	−0.523	0.00	−0.371	−0.256	0.573
Rape	−0.355	0.307	−0.120	0.793	0.167	0.281	−0.171
Robbery	−0.395	−0.350	−0.242	−0.377	0.00	0.697	−0.178
Assault	−0.425	−0.261	−0.100	0.00	0.285	−0.606	−0.537
Burglary	−0.393	0.261	0.429	−0.115	−0.717	0.00	−0.257
Theft	−0.307	0.717	0.00	−0.438	0.379	0.00	0.222
Vehicle	−0.359	−0.328	0.675	0.100	0.301	0.00	0.455

The first two principal components can be used to construct a scatter plot of the data (see Figure 10.3).

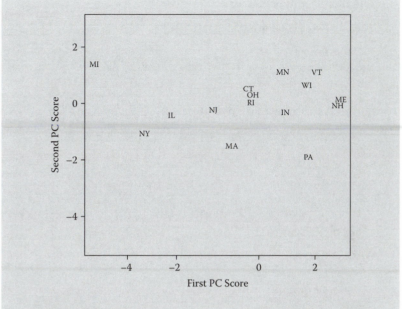

FIGURE 10.3   Plot of the cities in the space of the first two principal components.

10.6 SUMMARY

Multivariate data matrix: $\mathbf{X} = \begin{pmatrix} x_{11} & \cdots & x_{1p} \\ \vdots & \ddots & \vdots \\ x_{n1} & \cdots & x_{np} \end{pmatrix}$

Covariance matrix: $\boldsymbol{\Sigma} = \begin{pmatrix} \sigma_{11} & \sigma_{12} & \cdots & \sigma_{1p} \\ \sigma_{21} & \sigma_{22} & \cdots & \sigma_{2p} \\ \vdots & \vdots & \vdots & \vdots \\ \sigma_{p1} & \sigma_{p2} & \cdots & \sigma_{pp} \end{pmatrix}$

Sample covariance matrix: $\mathbf{S} = \sum_{i=1}^{n} (\mathbf{x}_i - \bar{\mathbf{x}})(\mathbf{x}_i - \bar{\mathbf{x}})' / (n-1)$

Sample correlation matrix: $\mathbf{R} = \mathbf{D}^{-1/2}\mathbf{SD}^{-1/2}$

Multinomial distribution:

$$\Pr(n_1, n_2, \ldots, n_k) = \frac{n!}{n_1! n_2! \ldots n_k!} p_1^{n_1} p_2^{n_2} \ldots p_k^{n_k}, n_i = 0,1,\ldots,n; n = \sum_{i=1}^{k} n_i$$

Multivariate normal distribution:

$$f(\mathbf{X}) = \frac{1}{(2\pi)^{p/2}} |\boldsymbol{\Sigma}|^{-1/2} \exp\left\{-\frac{1}{2}(\mathbf{X} - \boldsymbol{\mu})\boldsymbol{\Sigma}^{-1}(\mathbf{X} - \boldsymbol{\mu})'\right\}; -\infty < X_i < \infty, i = 1\ldots p$$

Linear combinations: If $L = \sum_{i=1}^{p} c_i X_i = \mathbf{c}'\mathbf{X}$ where $\mathbf{c}' = [c_1, c_2, \ldots, c_p]$ is a vector of scalars and if $\mathbf{X} \sim MVN(\boldsymbol{\mu}, \boldsymbol{\Sigma})$ then $L \sim N(\mu, \sigma^2)$ where $\mu = \mathbf{c}'\boldsymbol{\mu}$ and $\sigma^2 = \mathbf{c}'\boldsymbol{\Sigma}\mathbf{c}$.

Wishart distribution:

$$f(\mathbf{S}) = \frac{\left(\frac{1}{2}n\right)^{\frac{1}{2}p(n-1)} |\boldsymbol{\Sigma}^{-1}|^{\frac{1}{2}(n-1)} |\mathbf{S}|^{\frac{1}{2}(n-p-2)}}{\pi^{\frac{1}{4}p(p-1)} \prod_{j=1}^{p} \Gamma\left\{\frac{1}{2}(n-j)\right\}} \exp\left\{-\frac{1}{2}n\,\mathrm{trace}(\boldsymbol{\Sigma}^{-1}\mathbf{S})\right\}$$

Principal components transformation: $x \to y = \boldsymbol{\Gamma}'(x - \boldsymbol{\mu})$
Properties of principal components:

$$E(y_i) = 0, \mathrm{Var}(y_i) = \lambda_i, \; \mathrm{Cov}(y_i, y_j) = 0, i \neq j, \mathrm{Var}(y_1) \geq \mathrm{Var}(y_2) \ldots \geq \mathrm{Var}(y_p) \geq 0,$$

$$\sum_{i=1}^{p} \mathrm{Var}(y_i) = \mathrm{trace}(\boldsymbol{\Sigma}), \prod_{i=1}^{p} \mathrm{Var}(y_i) = |\boldsymbol{\Sigma}|.$$

SUGGESTED READING

Everitt, BS and Hothorn, T (2011) *An Introduction to Applied Multivariate Analysis with R*, Springer, New York.

Morrison, DF (1990) *Multivariate Statistical Analysis*, McGraw-Hill, New York.

Timm, NH (2002) *Applied Multivariate Analysis*, Springer, New York.

Index

.